Nondestructive Evaluation for Aerospace Requirements

Nondestructive Testing Monographs and Tracts

A series edited by Warren J. McGonnagle, *Elmhurst, Illinois*

1. **Microwave Nondestructive Testing Methods**
   by Alfred J. Bahr

2. **Acoustic Emission**
   edited by James R. Matthews

3. **Electromagnetic Methods of Nondestructive Testing**
   edited by William Lord

4. **Automated Nondestructive Testing**
   edited by Warren J. McGonnagle

5. **Nondestructive Evaluation for Aerospace Requirements**
   edited by Gary L. Workman

Nondestructive Evaluation for Aerospace Requirements

Papers from a conference held
26–27 August, 1987
University of Alabama,
Huntsville

Edited by

Gary L. Workman
University of Alabama, Huntsville

GORDON AND BREACH SCIENCE PUBLISHERS
New York • London • Paris • Montreux • Tokyo • Melbourne

Gordon and Breach Science Publishers

Post Office Box 786
Cooper Station
New York, New York 10276
United States of America

Post Office Box 197
London WC2E 9PX
England

58, rue Lhomond
75005 Paris
France

Post Office Box 161
1820 Montreux 2
Switzerland

3-14-9, Okubo
Shinjuku-ku, Tokyo
Japan

Private Bag 8
Camberwell, Victoria 3124
Australia

Library of Congress Cataloging-in-Publication Data

Nondestructive evaluation for aerospace requirements.

(Nondestructive testing monographs and tracts ; v. 5)
Includes index.
1. Aerospace industries—Materials—Testing—Congresses. 2. Non-destructive testing—Congresses.
I. Workman, Gary L. II. Series.
TL698.N66 1989 629.134'2 88-24315
ISBN 2-88124-305-3 (Paris)

Contents

Preface to the Series

This series concentrates on the methods of, and techniques in, nondestructive testing (NDT) and will publish information and data obtained from applications, experimentation, and theory; improved methodology; instrumentation modification and developments; application of the methodology to new and different problems or problem areas, and analysis, discussion, and evaluation of standard methodology. Each volume may be devoted to one method or subject or may contain papers on a number of methods.

Members of the Editorial Board will assist in the selection of the subject matter, as well as in the evaluation of manuscripts submitted for publication in the series.

We believe that these monographs and tracts will help advance the science, engineering, and technology of NDT and stimulate its application to industrial problems, thus helping to ensure the quality, reliability, and safety of industrial materials, components, and systems. Some of the volumes will also be useful in seminars and courses in NDT materials and related areas.

Comments concerning this series are welcome.

Warren J. McGonnagle

Preface

The papers contained in this monograph were selected from a much broader agenda given at a conference on "Nondestructive Evaluation for Aerospace Requirements," which was held in Huntsville, Alabama on August 26 and 27, 1987. The objective of the conference was to provide a forum for discussion of the rapidly changing requirements in NDE resulting from the many technological changes occurring in structures and propulsion systems used for aerospace transportation, space applications, and missile systems. With the advent of new materials and new joining techniques, we find ourselves striving continuously to improve the research and development of new NDE techniques as well as to promote better quantification of the more traditional ones. In adapting NDE to these new technologies, we also provide for a better understanding of the inherent limitations of these advancements.

Too often, NDE becomes the last stage in the development of a new product or deployable system much too late in the manufacturing development cycle to achieve properly what should have been an easily-achieved goal. What is missing in these programs is the feedback between the designers and the NDE specialists. Innovative enhancements in our ability to determine the integrity and reliability of new systems can only occur through long-term relationships between researchers, designers, and the fabricators. NDE, on the ground or in flight, is a necessity for both manned and unmanned applications. We will always have the requirement for NDE, whether it exists as an on-line monitor of process reliability or used for final systems verification. These discussions occurred throughout the conference, indicating the need for such a forum to discuss key issues which will impact developments for the next generation of aerospace systems. We hope that they will establish a baseline for successful implementation of NDE in future systems. Finally, we trust that the readers of this monograph will concur with the conference attendees in the importance of including NDE throughout the design and development phases in all future aerospace systems.

We wish to thank the University of Alabama in Huntsville for hosting the "Nondestructive Evaluation for Aerospace Requirements" conference as well as the supporting societies for helping to spread the word. A special thanks should go to all the speakers and session chairpersons who made it happen.

Gary L. Workman

Advanced NDE for Solid Rocket Motors

Joseph S. Heyman

NASA, Langley Research Center
Hampton, Virginia 23665

ABSTRACT

Recent NASA-LaRC advances in NDE for Solid Rocket Motors (SRM) will be discussed. The synthesis of these technologies was initiated through the development of physical models of the SRM for the type of probing energy used to investigate the integrity of the motors. Both ultrasonic and thermal energy models were used to optimize tests of buried SRM delaminations. In each case, the models resulted in experimental techniques that were improvements over standard methodology.

INTRODUCTION

The importance of quantitative NDE to mission assurance is impacting many SRM programs today. It is not enough to "NDE'" an item—one must understand critical failure mechanisms in a given design and test for system integrity based on a thorough knowledge of the system in question. The implication of that statement changes the nature of NDE in a significant fashion. In one bold stroke, NDE must take a parallel position to engineering designs such as stress, thermal, fatigue and fracture, etc. As an example, use of finite element analysis has allowed design engineers to "fly" a structure on a computer without building a single element to assess a particular concept. Of course, promising models require verification and experimental testing. Today, NDE must be raised to a similar professional level.

A first approach to a new design should include design for inspection with the same dedication as shown by the engineers to design for intended stresses. That goal, unfortunately, has not been achieved on a broad scale. One of the reasons NDE has not been elevated to the parallel professions, has been the annecdotal history of our trade. We have all participated in discussions where a statement of NDE is made such as "that part looks OK!" In contrast, one may ask a designer what stress level a part has been designed for, and get a numerical answer. For NDE to be accepted as one of the design sciences, it must become a measurement science.

In this paper, we present two applications for NDE that were developed from models of the propagation of energy in complex structures representing SRM geometries. In each case, a normal approach to the NDE application did not reveal the required information and would have resulted in establishing a standard that did not represent the best information available from an inspection. A more careful analysis of the physical models, dictated technologies that improved the ultimate resolution and evaluation of the observed measurements.

ULTRASONIC EVALUATION OF BURIED SRM BONDLINES

A conventional application of ultrasonics to the SRM involves using broadband pulse-echo to insonify the material. Normal analysis of the returning signals consists of setting a time gate of interest and a signal threshold level established from test parts with defects. Such tests were made on samples simulating the Shuttle SRM internal bondlines between the fuel, liner, insulation, bond layer and steel case. Case/insulation disbonds were easily detected while disbonds between the liner and the fuel were difficult to detect with certainty. Multiple ringing of the ultrasonic pulse in the low-loss steel case obscured energy coming from the buried bonds.

An ultrasonic numerical model of the SRM structure was developed[1] to evaluate performance of unconventional approaches. The model insonified the structure with a long tone burst[2] plane wave normal to the steel surface. A transmission line impedance was calculated including the effects of each reflecting interface in the structure. The resulting reflected wave consists of the superposition of all reflected waves and was examined as a function of frequency. The values used for the model are shown in Table 1.

Figure 1 shows the calculated spectral response of the SRM viewed from the steel case side as a function of frequency. The two curves represent a numerical evaluation of a good bond and an unbond, between the fuel and the liner. The figure shows small differences between the two bond conditions with the average difference decreasing with frequency. The decrease is caused by the high attenuation of the insulation and liner material preventing ultrasonic energy from penetrating into the

TABLE 1- Ultrasonic values used in the model.

Material	Velocity m/sec	Density kg/m^3	Attenuation nepers/m
Steel	6080	7600	0
Insulation	1810	1218	$1.11 \times 10^{-5} f^{1.225}$
Liner	1500	999	$23.0 f^0$
Fuel	2170	1503	$213. f^0$

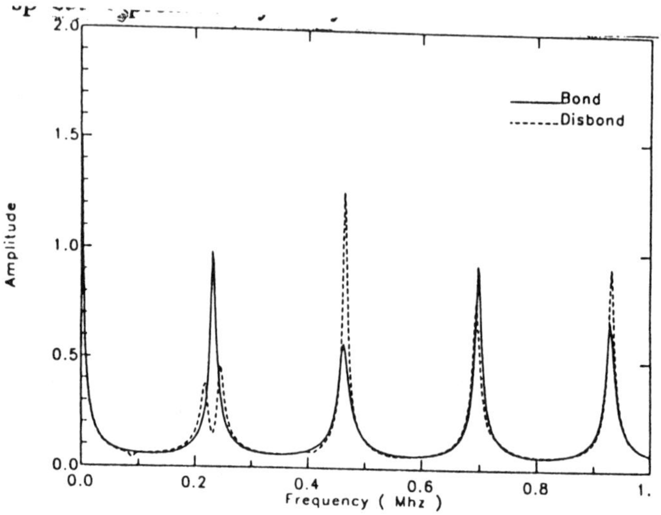

FIGURE 1- Calculated spectral response of the SRM for two conditions representing good bond and lack of bond between the fuel and the liner.

deeper structures. The difference between the two curves depends strongly on frequency, with maximum contrast (difference between the bond and unbond condition) occurring at resonance peaks spaced approximately every 240 kHz.

The difficulty in examining the deeper bondline with pulse-echo is now clear. Pulse-echo uses a narrow pulse which comprises many frequencies. Only those frequencies which lie in the narrow range of a mechanical resonance contribute to the measurement of the disbond. The other frequencies represent background noise through which the weaker bondline signals must be discriminated.

The results of the numerical calculations point to the benefit of using a narrow band of ultrasonic frequencies near the composite resonance of the structure involved.

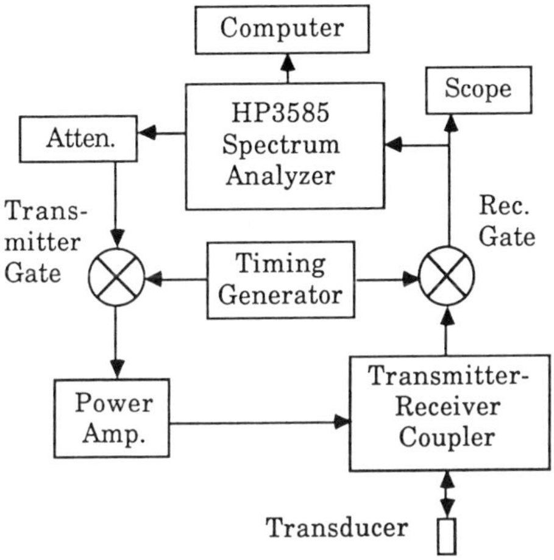

FIGURE 2- Block diagram of the experimental test set-up used to obtain ultrasonic spectral information on the SRM.

The resonance of choice would be the one near 0.2 MHz, where the bonded signal is nearly twice the unbonded signal, or near 0.5 MHz where the opposite is true. To verify the model, we built a sample consisting of a flat steel plate with multiple layers of a rubber material similiar to the insulation/liner followed by either an unbond (an air gap) or a bond (coupling directly to water). The thicknesses were choosen to simulate the SRM itself.

A block diagram of the experimental test is shown in fig. 2. A spectrum analyzer measured the response of the sample. The output from a tracking generator in the spectrum analyzer was gated (40 microseconds) and amplified to drive the insonifying 0.5 MHz, 3.2 cm diameter flat transducer which was coupled to the sample through a 70 microseconds water delay line. The reflected wave was time gated into the spectrum analyzer at a time that eliminated the reflection interface between the steel and the water delay line. The spectral response of the sample is shown in fig. 3 which is in excellent agreement with the calculated response.

A more realistic sample of the SRM was fabricated by Morton Thiokol using the same materials, processes and thickness as are used in the SRM. The sample was 30.5 cm (12 in.) square with the 3.66 m diameter (12 feet) radius of curvature as found in the motor steel case. A brass wedge shaped shim was cured into the sample at the liner/fuel interface. The wedge extended from the sample and was pulled after curing to create a disbond in the desired location. The shape of the sample is shown in fig. 4.

An image of the sample was made using the spectrum analyzer system scanning the sample with a step size of 0.2 cm with the image area 8 × 12 cm. A similar scan was made using a conventional pulse-echo system gating the signal for the liner/fuel interface using peak detection and various threshold levels. The conventional scan could not resolve the disbond above background variations in the signal. The image obtained with the tone burst

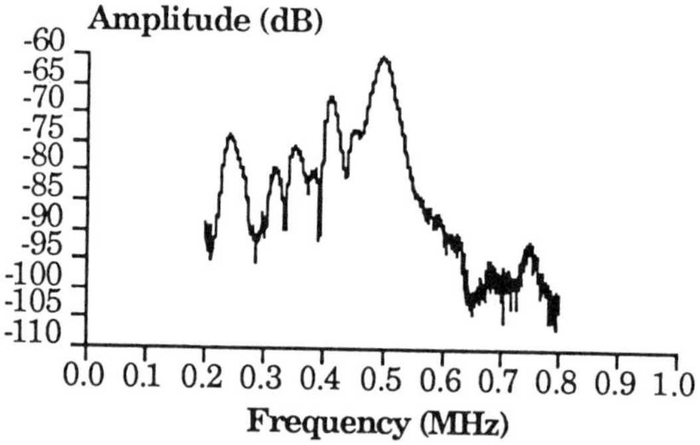

FIGURE 3- Spectral response of the SRM sample showing the computation model resonance peaks. Detection of internal disbonds is optimized using ultrasonic frequencies corresponding to the high contrast indicated in fig. 1.

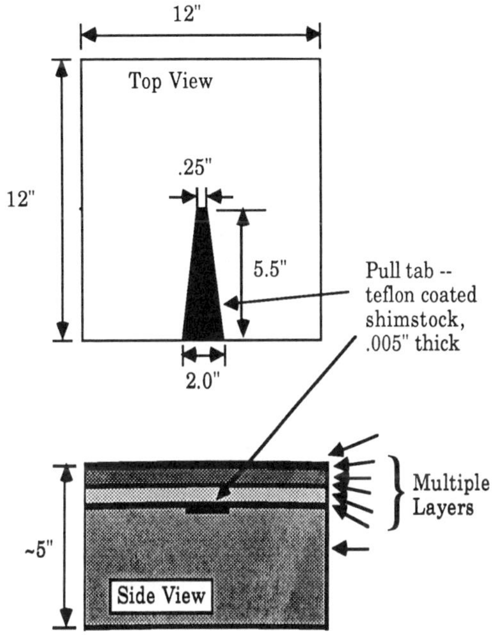

FIGURE 4- SRM samples with built-in disponds to simulate possible manufacturing defects. Disbonds were placed in various bondlines to evaluate their detection.

system of fig. 2 clearly resolved the wedge delamination shown in fig. 5.

The improved signal to noise was made possible by using the physical model information. A practical scan system is currently under development at LaRC to lock an ultrasonic system to the SRM resonance peak and track any changes in that frequency caused by variations in the layered structure. The returning signal strength is also monitored, forming a pair of measurements representing both amplitude and phase information for improved reliability.

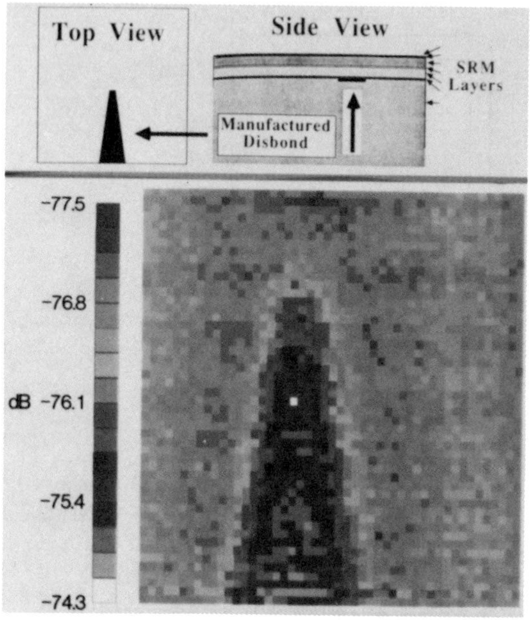

FIGURE 5- Grey scale ultrasonic image of reflection amplitude at the resonance frequency (240 kHz) of the SRM motor. Image shows the wedge disbond between the fuel and the liner.

THERMAL EVALUATION OF BURIED SRM BONDLINES

Although ultrasonics is a logical approach to examine the SRM, it is not the only technology appropriate for the motor. Different, complimentary techniques add significant reliability factors to the inspection. In addition, for thermal energy, there may be significant cost benefits for large area rapid scanning. Traditional thermal NDE consists of surface heating while measuring the temperature radiometrically with an IR camera. A near surface disbond would appear as a transient hot spot.

We have developed OPTITHERMS (OPTIcal THermal InfraRed Measurement System),[3] a thermal NDE system composed of a scanned modulated laser, an IR camera, an image frame grabber, and most important, a thermal model. Similar to ultrasonics, the model is used to interpret the data and to point the way to improved signal-to-noise measurements.

The thermal dynamics are simulated using an equivilent circuit analysis where voltage is temperature difference, resistance is related to diffusion (thermal impedance) and capacitance to heat capacity. The different layers of the motor are represented by a series/parallel circuit of the proper elements with values determined to map the thermal diffusion physics into the equivalent electrical analogues.[4] Figure 6 shows the general model with the

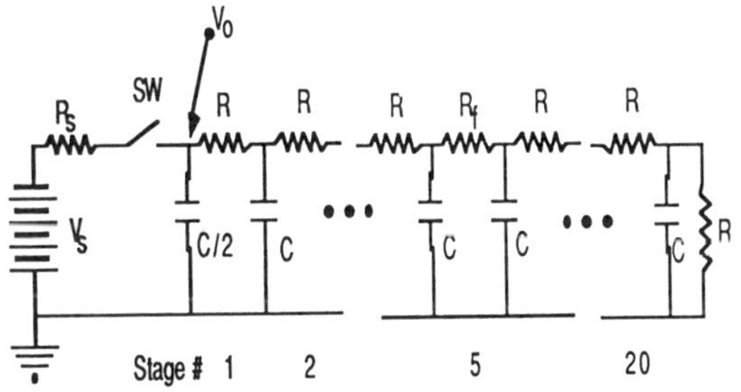

FIGURE 6- Equivalent electrical model of the thermal properties of the SRM in the inhibitor region. The switch/battery with internal resistance R_s represents the thermal source while R_f is a flaw. The rest of the model corresponds to internal elements in the SRM with V_0 expressing the surface temperature. The switch closes at time = 0.

switch/battery representing the input heat source with its source impedance and thermal difference.

The results of the analysis of this straightforward model show that the best signal-to-noise one can achieve is equal to the highest contrast defined as the maximum difference in voltages, V_0, for the case with and without a flaw. The time evolution of that voltage difference (difference in surface temperature) is shown in fig. 7 for two cases of surface impedance.

The curve labled "source pulsed" represents heating/cooling the inhibitor region of the SRM which is in contact

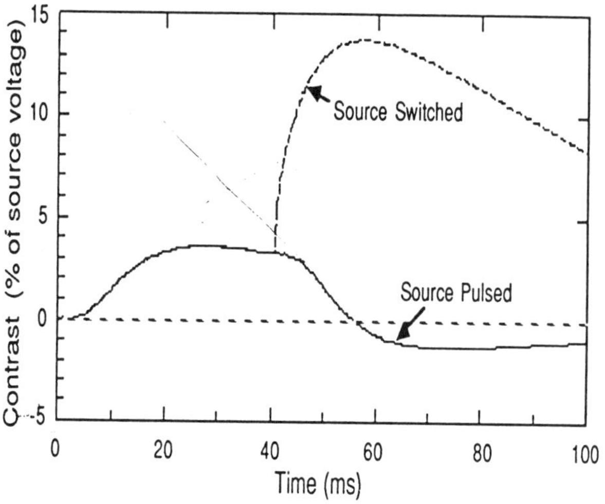

FIGURE 7- Computational results from the model of fig. 6. The results show the effect of the water bath heating and acting as a low surface impedance. Source pulsed corresponds to alternate flow of hot and cold water in the bath while source switched corresponds to removing the hot bath after the soak time. The contrast is the surface temperature difference with and without a flaw.

with a liner and the fuel. The heat is applied with a hot water bath in contact with the sample while the temperature is monitored on the sample under the water bath. The curve labled "source switched" represents opening the switch (SW) thus changing the surface impedance. Such a heating condition is simulated by placing the hot water bath in contact with the sample, and then lifting that bath off the sample at 30 seconds time (40 milliseconds in the electrical model). The increase in contrast for the water bath technique shows that the source imped-

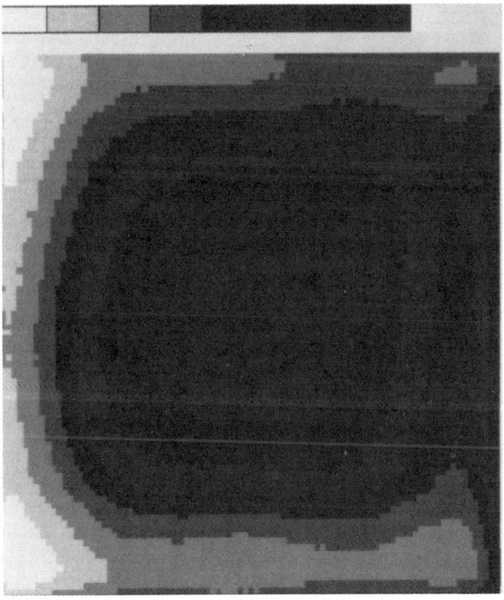

FIGURE 8a- Surface temperature profile of the SRM steel case after uniform surface heating followed by a cooling time period.

ance, related to the water contact with the sample, can play a significant role in signal-to-noise optimization.

A more complex analysis of the thermal system involves the determination of equivalent thermal sources and sinks that effect the flow of thermal energy. This is similar to saying that we do not want to measure voltages since voltage differences may be small. Instead, we wish to measure current flow into or out of the sample. An example of the power of this approach is shown in fig. 8a, where the image shows the temperature of the surface

FIGURE 8b- Thermal NDE image showing the wedge disbond. The image is the second spacial derivative of the temperature and represents the flaw as a thermal source.

of the SRM sample. Note that the main visible feature is the effect of heat loss at the edges producing a central peak in temperature with edge cooling dominating the image.

Figure 8b, on the other hand, was obtained from the second spacial derivative of this image to extract thermal sources. The wedge shaped delamination clearly stands out from the background. A delamination appears as a heat source since it can provide more energy to the surface than the surrounding material. By imaging the processed data as a heat source, we are able to differentiate delaminations from other processes in the sample thus improving our ability to interpret the image.

CONCLUSION

The analysis of ultrasonic and thermal physical models of the SRM have resulted in test proceedures with improved buried defect detectability for the SRM bondlines. In each case, the normal NDE test would not have been able to characterize or locate the defect in the tested structure. Many labs are now applying similar physical analysis to raise the professional level of NDE to parallel other engineering and scientific disciplines.

It is important that program management embrace these new thrusts and include NDE considerations at project initiation. "Design for inspection" should be part of any criteria spelled out in contract specifications with the same emphasis as is placed on "'design for stress." The significance of these changes will be improved safety, reliability, adherence to schedule, and cost savings. A side benefit will be the assurance of project managers that what is measured, is quantifiable—or what you see is what you get!

References

1. Madaras, Eric I., Winfree, William P., Smith, B. T. and Heyman, Joseph S., "Detection of Bondline Delaminations in Multilayer Structures wth Lossy Components," *1987 IEEE Ultrasonics Symposium Proceedings*, 1987 (in publication).
2. Cantrell, John H.; Heyman, Joseph S.: "Ultrasonic Spectrum Analysis Using Frequency-Tracked Gated RF Pulses," Acoust. Soc. Am. *67*, 1623–1628 (1980)
3. Heath, D. M., Welch, C. S., Winfree, W. P., Heyman, J. S. and Miller, W. E., "Quantitative Thermal Diffusivity Measurements of Composites," *Review of Progress in Quantitative Nondestructive Evaluation*, Vol 5B, D. O. Thompson and Dale E. Chimenti, eds, 1986, pp 1125–1132.
4. Welch, C. S. "Contrast Enhancement Through Source Impedance Control In Thermal NDE," *Review of Progress in Quantitative Nondestructive Evaluation*, Williamsburg, VA 1987, (in press).

Ultrasonic Investigations on Contacting Interfaces—Applications to Fracture Mechanics

O. Buck*, R. B. Thompson*, D. K. Rehbein*, and D. D. Palmer†

* *Center for Nondestructive Evaluation and Ames Laboratory, USDOE Iowa State University, Ames, Iowa 50011*
† *McDonnell Douglas Corporation, P.O. Box 516, St. Louis, Missouri 63166*

ABSTRACT

Contacting interfaces exist under a variety of conditions. During fatigue crack growth, e.g., the crack closes partially due to the compressive stresses set up by the deformed material. It has been established that this crack closure can have pronounced effects on the rate of crack propagation and thus on fatigue life. A diffusion bond is a second example of a contacting interface. Since diffusion bonding is now under serious consideration as a means to join components, the question is, "How good is the bond in terms of failure strength?" Another longstanding problem exists at sliding interfaces, which in all cases leads to friction and in some cases to wear.

Our research has concentrated recently on the first two types of interfaces where the contact along the interfaces usually occurs by asperities under compression or through connecting ligaments. This paper deals with the mechanical effects of such contacts. Asperity contact along the fracture surface of a fatigue crack shields the crack tip, in part, from the externally applied driving force. Using information from acoustic experiments, we estimated the geometry of the asperities. These data provide the contacting stress as well as the shielding stress intensity factor. Acoustically, a diffusion bonded interface looks very similar to that of a partially closed crack. In this particular case the acoustically determined geometry of well bonded ligaments can be verified by fractography of destructively tested samples whose bond strengths have also been determined. For good diffusion bonds, the bond strength may be predicted from the ligament geometry and the threshold stress intensity of the material.

INTRODUCTION

An accurate determination of crack length is a primary prerequisite for fracture toughness and fatigue crack propagation measurements. Besides standard optical techniques there are a number of methods available now which are capable of performing such measurements, such as compliance gauge, electrical potential drop, and acoustic techniques.[1,2] In applying any one of these techniques to determine the crack length it becomes immediately apparent, however, that the measurement has to be performed at an applied load such that the crack is fully open. At lower loads, the crack will close down on itself, leading to erroneous crack length measurements. On the other hand, crack closure can have a significant effect on fatigue crack propagation since closure shields the crack tip in part from the applied driving force on the crack.[3] At present the lowest stress intensity factor, K_{cl}, at which the crack is still fully open, is the only information derived from such measurements in addition to the above mentioned crack length measurement. This then leads to an "effective" stress intensity range[3]

$$\Delta K_{eff} = K_{max} - K_{cl} \tag{1}$$

in contrast to the usually used alternating stress intensity range

$$\Delta K = K_{max} - K_{min} \tag{2}$$

which does not include crack closure effects. In specific cases, use of Eq. (1) has led to unique relations between the crack propagation rate, da/dN, and ΔK_{eff}, independent of stress overload effects or of the applied stress ratio.[4] Unfortunately, the precise determination of K_{cl} is difficult as well as controversial.[4] As discussed in the following, K_{cl} is not even the correct quantity to be used for the determination of ΔK_{eff} since contact in the wake of the crack tip does not completely shut off the effect of the externally applied stress on the crack tip. Each contact point in the wake of the crack creates its own stress intensity factor. The sum of all stress intensities created by the contacts is called the shielding stress intensity factor. Since any stress intensity factor is geometry dependent, knowledge of the asperity contact geometry and of their location with respect to the crack tip is a primary prerequisite to determine shielding. As discussed earlier[5] and in the following, such knowledge can be obtained by acoustic techniques.

A similar problem occurs in bonding of similar or dissimilar materials, such as by diffusion bonding. In recent years, this process has found considerable usage in both the nuclear power and aerospace industries. This process requires the compression of mating surfaces at an elevated temperature for a given time. If optimum conditions of time, temperature, pressure and surface cleanliness are achieved, diffusion of material across the interface will occur, yielding interfacial mechanical properties identical to those of the bulk material. The use of insufficient bonding conditions may result in void formation, precipitation of undesired phases or lack of grain growth across the interface. The consequence will be an interface that is less than fully bonded, which will result in degradation of the mechanical properties. Applications of diffusion bonding to nuclear reactor fuel elements, helicopter rotor hubs, jet engine turbine blades, etc., thus make the ability to characterize the strength of these interfaces highly desirable.

Recent experimental results have shown that the ultrasonic reflectivity and transmissivity of an interface can be used to investigate details of the interface. In one application,[6] such measurements have been used to determine the radial stress component in a shrink-fit coupler and the weld quality in a pinch welded tube.[7] Regalbuto[8] applied pulse-echo measurements to the study of diffusion bonded joints with some initial success.

On the other hand, a number of theoretical models have been developed with the objective being to quantify the transmissivity and reflectivity of acoustic waves at partially contacting interfaces.[9–11] These models attribute these signals from the interface to asperities in contact between the two surfaces. The present paper describes briefly the applications of these models to the interpretation of experimental results obtained on fatigue cracks and diffusion bonds.

EXPERIMENTS

Figure 1 schematically illustrates two possible arrangements for experimental ultrasonic investigation of partially contacting interfaces. Both arrangements interrogate the interface and yield results that contain information on the amount of contact and size of the individual contacts at the interface. In both cases, broadband focused acoustic transducers (2–15 MHz) are used, with the focus being in the plane of the fatigue crack or the diffusion bond, respectively. During the experiments the specimens are translated with respect to the transducers to interrogate different regions of the contacting areas. The experiments are special cases of more general ones in which the angles of incidence and observation may be arbitrarily chosen to observe either transmitted or reflected waves or waves diffracted and mode converted at the contact points.

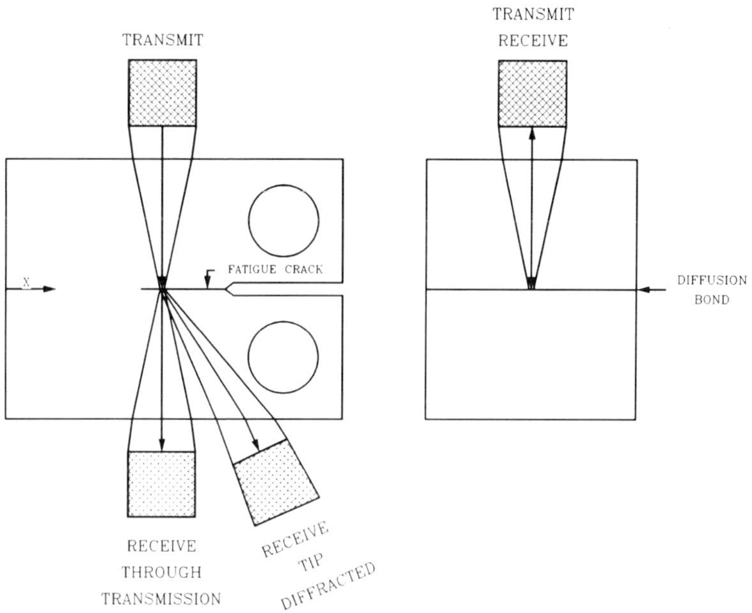

FIGURE 1- Arrangements for experimental ultrasonic investigation of partially contacting interfaces, transmission and diffraction through fatigue crack (left) and reflection from diffusion bond (right).

Figure 2 schematically illustrates the situation that the incident acoustic beam sees. The contacting asperities form a layer that is less stiff than the bulk of the material. The figure shows the asperity arrangement for a diffusion bond at the top and a fatigue crack at the bottom. For both cases, it is the objective of the experiments to determine the distance C between asperities as well as their diameter d.

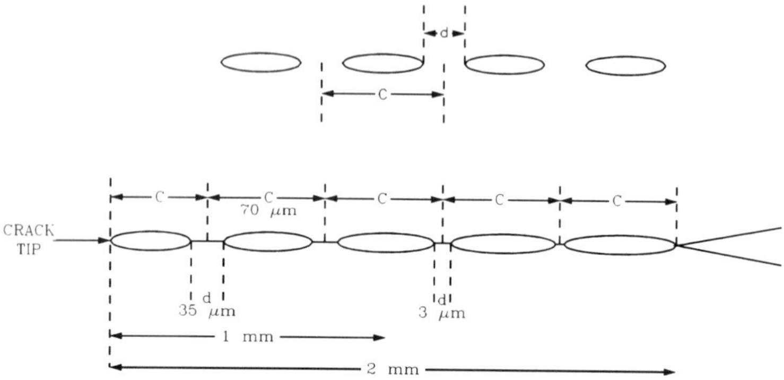

FIGURE 2- Schematic illustration of asperity arrangement at diffusion bond interface (top) and fatigue crack (bottom).

Crack Tip Shielding

The effect of the closure region on the ultrasonic response of the crack in through transmission is illustrated in the top part of fig. 3. For a fully open crack the transition between the uncracked ligament and open crack would be much sharper with all curves crossing at a transmission coefficient of 0.5.[12] The closure region broadens this transition and causes the cross-over points to be shifted, effectively blurring the exact location of the crack tip.

Calculation of the acoustic response of such a crack is possible using the distributed spring model.[10,13] This model is based on the change in stiffness of the interface layer, κ, due to the partial contact of the layer, where κ

FIGURE 3- Ultrasonic response of partially contacting fatigue crack. (Top) Experimental results. (Bottom) Distributed spring model prediction. Value shown on each curve is frequency in MHz.

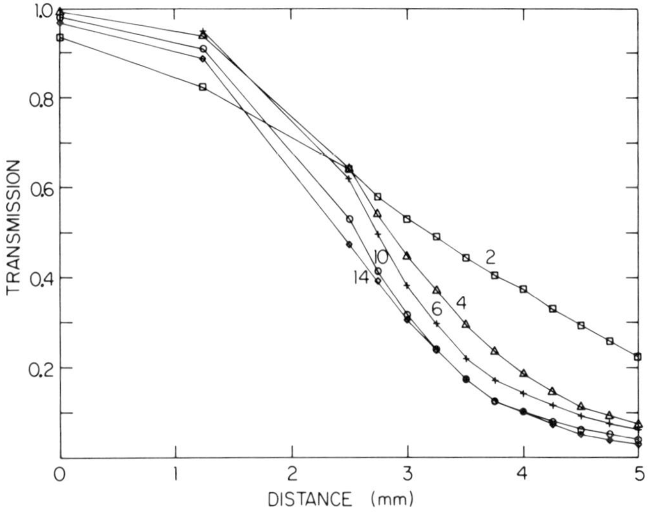
1.0
0.8
0.6
0.4
0.2
TRANSMISSION
0
1
2
3
4
5
DISTANCE (mm)
2
4
6
10
14

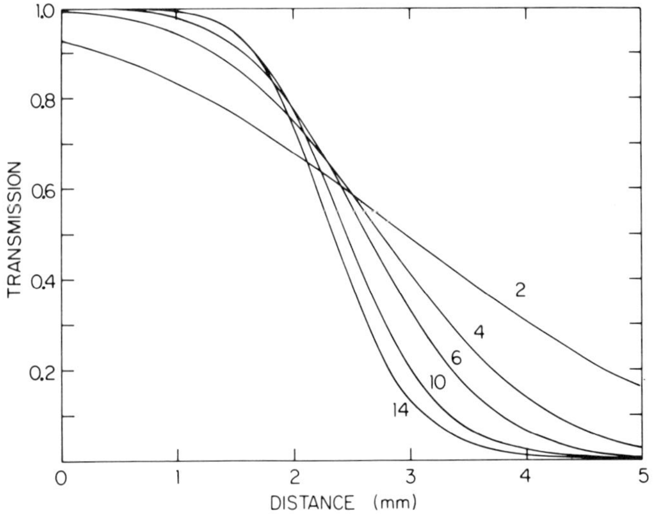
1.0
0.8
0.6
0.4
0.2
TRANSMISSION
0
1
2
3
4
5
DISTANCE (mm)
2
4
6
10
14

depends on C and d. Assuming that the stiffness of the layer can be represented by

$$\kappa(x) = \kappa_0 e^{-\beta x} \tag{3}$$

the theoretically expected acoustic response can be evaluated, as shown in fig. 3, bottom, and matched to the experimental results. Including diffraction results, comparison of theory and experiment then provides information on the density of (or average distance between) contacts as well as their size.[13] For a fatigue crack in Al 7075-T651, grown under constant ΔK conditions, the total length of the closure region was found to be about 2 mm. The information on the average distance between contacts and their size can now be used to calculate the crack tip shielding stress intensity factor, K_{sh}.

Consider a crack which is initially loaded in the fully open condition. As the load decreases, the stress intensity factor assumes the value K_{cl} as the first row of individual asperities (shown on the left side of fig. 4) contacts the

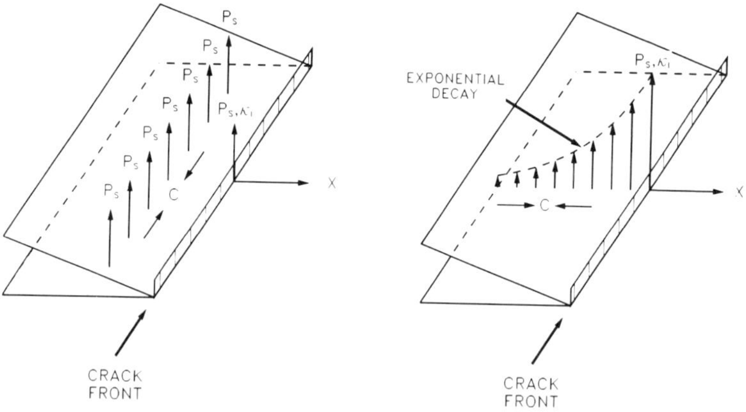

FIGURE 4- Illustration of load carried by contacting asperities. (Left) First row parallel to crack front. (Right) Successive contacts perpendicular to crack front.

opposite fracture surface. The average distance between contacting asperities, C, does not change on further unloading; the average diameter of the asperities, however, increases, as does the load, P_s, carried by each individual contact. As unloading continues additional rows of asperities come into contact with the opposite fracture surface (shown on the right side of fig. 4). Each individual asperity which comes in contact with the opposite crack surface produces a stress intensity factor, K_i. The body containing the crack is assumed to be infinite in the z-direction and therefore K_i is valid for plane strain conditions. Superposition of the stress intensity factors of a series of such individual contact points, assuming that the distance between two neighboring contact points is C, yields the stress intensity factor contribution[13]

$$dK_{sh} = \left(\frac{2}{\pi}\right)^{1/2} \frac{\Delta P}{BC^{1/2}} \tag{4}$$

which is identical to results obtained by Beevers et al.[14] for a series of infinite contacting strips parallel to the crack tip. B in Eq. (4) is the specimen thickness and ΔP the load carried by the strip.

On average, the load ΔP carried by the strip can be related to the static stress across the partially closed crack, σ_0, as

$$\Delta P = dP = B\sigma_0 dx \tag{5}$$

where x is a coordinate perpendicular to the crack front (see fig. 4). Based on earlier work by Haines[9] and Thompson et al.[10,15] connecting the average static stress, σ_0, to the acoustic transmission coefficient, it has been shown[16] that

$$\sigma_0 = \left(\frac{\kappa}{k^*E}\right)^2 \left(\frac{\pi}{N}\right) P_m \tag{6}$$

where κ is the interface stiffness, defined in Eq. (3), $k^* = 2$, E the elastic modulus, N the areal contact density and P_m is the "flow pressure" of the material (usually three times the ultimate tensile strength). Assuming N to be a constant, which is determined by factors related to microstructure and the mechanisms producing closure, than all the quantities in Eq. (6) are independent of x, except for κ. When Eq. (3) is valid, one can easily calculate the contribution dK_{sh} of contacting strips of width dx to the total stress intensity factor.

Under the constraint that the separation between strips is much less than β^{-1}, the total contribution of the closure region will be

$$\int dK_{sh} = K_{sh} = \pi \left(\frac{\kappa_0}{k^*E}\right)^2 \frac{P_m}{N\beta^{1/2}} \tag{7}$$

For the unloaded compact tension specimen, the acoustically determined parameters κ and N can now be used to evaluate the shielding stress intensity factor K_{sh}, given by Eq. (7), as well as the static compressive stress σ_0 given by Eq. (6). For a crack in Al 7075-T651, with a yield stress of 500 MPa, the following values have been obtained[13] from Eq. (7):

$$K_{sh} \approx 6.8 \text{ MPa m}^{1/2} \ (\approx 7.5 \text{ ksi} \sqrt{\text{in}})$$

and for the compressive static stress at the crack tip ($\kappa = \kappa_0$) from Eq. (6):

$$\sigma_{0max} = 340 \text{ MPa} \ (\approx 50 \text{ ksi})$$

and therefore equal to about 70% of the yield stress. The estimated σ_{0max} is in agreement, relative to the yield stress, with x-ray diffraction measurements[17] of the residual stress in 4140 steel near the crack tip. In both cases, the residual stress drops off exponentially with distance away from the crack tip. Since the estimated K_{sh} is a quantity integrated over the total closure regime, it is thought, therefore, that the estimate of K_{sh} given above should be close to its true value.

It was already pointed out by Beevers et al.[14] that K_{sh} is a stress intensity factor which acts upon the crack tip as soon as the asperities come into contact at K_{cl}, as shown schematically in fig. 5. Above K_{cl} there is no asperity contact; thus, the stress intensity factor on the crack tip is totally supplied by the external load Q. Below K_{cl}, K_{sh} increases monotonically and is a maximum for $Q = 0$, when compressive loads on the specimen are disregarded. Thus it is actually the total K_{sh}, as given by Eq. (7), which determines the full shielding of the crack tip. Therefore, strain gauge and electrical potential drop measurements can only be considered as qualitative indicators for trends in the crack closure phenomena unless their changes during closure or opening of the crack can be interpreted in similar terms as in the present acoustic measurements. The present discussions should have

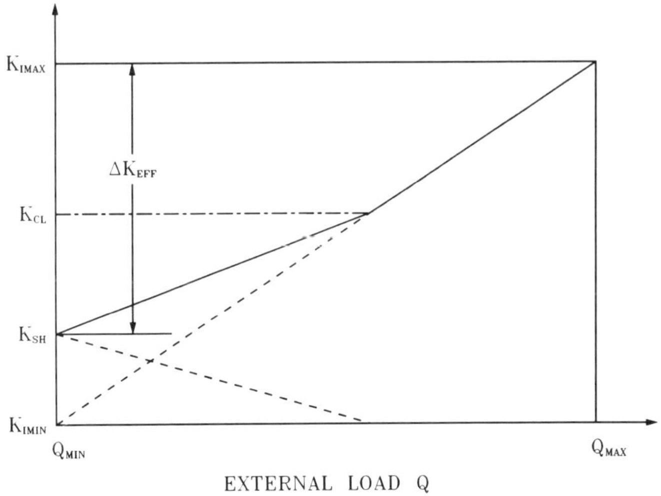

FIGURE 5- The effect of closure as a function of applied load on crack tip shielding.

made it clear, however, that if the full shielding effect is taken into account, the "effective" stress intensity range should be considered as

$$\Delta K_{eff} = K_{max} - K_{sh} \tag{8}$$

replacing Eq. (1).

Diffusion Bond Evaluation

Copper disks were diffusion bonded under constant pressure with variable parameters of time ($\frac{1}{4}$ h to 4 h) and temperature (400°C to 600°C) in an atmosphere of flowing hydrogen gas. Thus a series of diffusion bonds of varying quality was obtained. Using normal incidence longitudinal waves the reflection coefficient R of longitudinal waves reflected from the bond interface was determined. The samples were then cut into tensile specimens and strained to failure with the bond strength designated as the maximum stress that was placed on the sample during the tensile test. A graph of the results obtained is shown in fig. 6, including the ultimate stress obtained from a bulk specimen (not diffusion bonded) with a reflection coefficient of R = 0. As may be seen, the bond strength decreases sharply with increasing reflection coefficient.

After these destructive tests, optical micrographs of the failed diffusion bonds were obtained. From these micrographs, parameters such as the fraction of bonded area, as well as C and d, defined in fig. 2, were determined. For a fatigue crack, the information on C and d is sufficient to determine the stiffness of the interface layer, κ, which can now be correlated with the experimentally determined reflection coefficient. On the other hand, this spring model states that the acoustic reflection, R, from an imperfect interface is given by

$$R = \frac{(j\omega Z/2\kappa)}{(1+j\omega Z/2\kappa)} \tag{9}$$

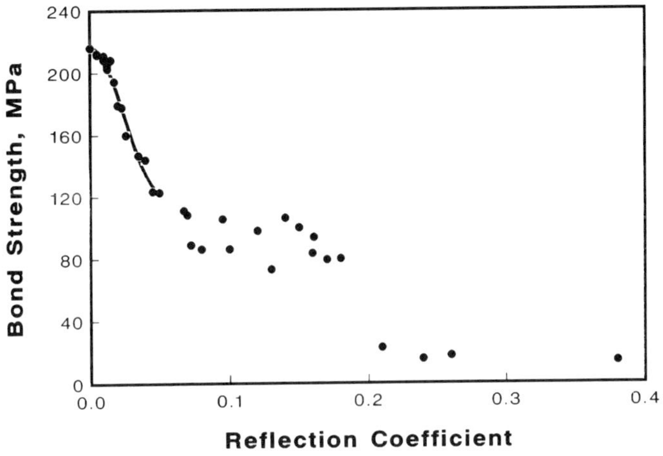

FIGURE 6- Correlation between bond strength and ultrasonic reflection coefficient for diffusion bonded samples.

where ω = angular frequency of the acoustic wave and Z = acoustic impedance (density times acoustic velocity in the metal). This equation thus also yields a relation between R and κ. For randomly distributed contacts or disbonds the experimental results are in very good agreement[18] with the theoretical prediction, obtained from Eq. (9), demonstrating the value of the spring model[10] to determine the geometry of the asperities from acoustic measurements. A more complete description of this work is given elsewhere.[18,19]

On the other hand, a single disbond at the interface, shown in fig. 2, may be treated as a crack-like object which can start to grow under an applied stress σ. Assuming that the disbonds are small and far apart, one may try to use a fracture mechanics approach using an approximation for penny-shaped cracks.[20] Under uniaxial tension these cracks would then start to grow and fill the specimen at a certain stress intensity factor for Mode I failure, K_I, which should be about equal to the threshold stress intensity factor K_{th}. Under these conditions,

$$\lim(d/C) \rightarrow 1 \tag{10}$$

one obtains from the disbond geometry that the bond strength σ should be

$$\sigma = \left(\frac{\pi}{2}\right)^{1/2} \frac{K_I}{\sqrt{d}} \tag{11}$$

or, in terms of the fraction of unbonded area,

$$\sigma = \left(\frac{\pi^{3/4}}{2}\right) \left(\frac{A_u}{A_0}\right)^{-1/4} \left(\frac{K_I}{\sqrt{C}}\right) \tag{12}$$

where A_u/A_0 is the fraction of unbonded area and C is as defined in fig. 2. From experimentally obtained results,[19] it was determined that good agreement between the measured bond strength σ and that calculated from Eq. (12) is achieved for the best bonds if K_I is about 0.9 $MPa\sqrt{m}$. On the other hand, Yeske and Weertman[21] determined measurable growth rates in copper at 2 $MPa\sqrt{m}$, so that the threshold level of crack growth must be somewhat lower and probably in the vicinity of 0.9 $MPa\sqrt{m}$.

It has been clearly established[19] that A_u/A_0 and C in Eq. (12) are simply functions of the reflection coefficient, R. If one is able to determine these parameters by completely nondestructive methods, then this approximation may prove to be a viable tool for a nondestructive bond strength determination.

CONCLUSIONS

In previous work, it has been shown that ultrasonic interrogation of a partially contacting interface can provide detailed information on the contact geometry at the interface, i.e., number and size of contacting areas. The discussions point out the possibility of obtaining all quantities necessary for life prediction and quantification of structural integrity in at least two primary areas, namely fatigue cracks and diffusion bonding.

The driving forces present at the tip of a crack can be determined thereby providing essential information about future crack propagation. Detailed knowledge of the contact geometry at a diffusion bond interface can yield the strength of the bonded structure. Further efforts are now under way in both areas of investigation.

Acknowledgment

This work was funded in part by the Center for NDE and in part by the Ames Laboratory which is operated for the U.S. Department of Energy by Iowa State University under Contract No. W-7405-ENG-82, supported by the Office of Basic Energy Science.

References

1. C. J. Beevers, ed., "The Measurement of Crack Length and Shape during Fracture and Fatigue," Engineering Materials Advisory Services, Cradley Heath, Warley, West Midlands, U.K., 1980.
2. C. J. Beevers, ed., "Advances in Crack Length Measurement," Engineering Materials Advisory Services, Cradley Heath, Warley, West Midlands, U.K., 1982.
3. W. Elber, in "Damage Tolerance in Aircraft Structures," ASTM STP 486, Am. Soc. Testing Materials, Philadelphia, 1971, p. 230.
4. J. C. Newman and W. Elber, eds., "Mechanics of Fatigue Crack Closure," ASTM STP 982, Am. Soc. Testing Materials, Philadelphia, 1988.
5. O. Buck, D. K. Rehbein, and R. B. Thompson, in "Effects of Load and Thermal Histories of Mechanical Behavior of

Materials," P. K. Liaw and T. Nicholas, eds., Metallurgical Society of AIME, Warrendale, PA, 1987, p. 49.
6. D. K. Rehbein, J. F. Smith, and D. O. Thompson, in "Review of Progress in Quantitative NDE, 3B," D. O. Thompson and D. E. Chimenti, eds., Plenum Press, New York and London, 1984, p. 1321.
7. D. K. Rehbein, D. K. Hsu, R. B. Thompson, and T. A. Jones, in "Review of Progress in Quantitative NDE, 6B," D. O. Thompson and D. E. Chimenti, eds., Plenum Press, New York and London, 1987, p. 1737.
8. J. A. Regalbuto, Matl. Eval., *30*, 66 (1972).
9. N. F. Haines, "The Theory of Sound Transmission and Reflection at Contacting Surfaces," Report RD/B/N4711, Central Electricity Generating Board, Research Division, Berkeley Nuclear Laboratories, Berkeley, England, 1980.
10. J-M. Baik and R. B. Thompson, J. Nondest. Eval., *4*, 177 (1984).
11. Y. C. Angel and J. D. Achenbach, J. Appl. Mech., *52*, 33 (1985).
12. R. B. Thompson and C. J. Fiedler, in "Review of Progress in Quantitative NDE, 3A," D. O. Thompson and D. E. Chimenti, eds., Plenum Press, New York and London, 1984, p. 207.
13. O. Buck, D. K. Rehbein, and R. B. Thompson, Engr. Fract. Mechanics *28*, 413 (1987).
14. C. J. Beevers, K. Bell, R. L. Carlson and E. A. Starke, Engr. Fract. Mechanics *19*, 93 (1984).
15. R. B. Thompson, B. J. Skillings, L. W. Zachary, L. W. Schmerr, and O. Buck, in "Review of Progress in Quantitative Nondestructive Evaluation, 2A," D. O. Thompson and D. E. Chimenti, eds., Plenum Press, New York and London, 1983, p. 325.
16. O. Buck, R. B. Thompson and D. K. Rehbein, J. Nondest. Eval. *4*, 203 (1984).
17. E. Welsch, D. Eifler, B. Scholtes, and E. Macherauch, in "Residual Stresses in Science and Technology," E. Macherauch and V. Hauk, eds., DGM Informations Gesellschaft Verlag, Oberursel FRG, 1987, p. 785.
18. D. D. Palmer, C. D. Roberts, D. K. Rehbein, J. F. Smith and O. Buck, in "Review of Progress in Quantitative NDE,

7B," D. O. Thompson and D. E. Chimenti, eds., Plenum Press, New York and London, 1988, p. 1335.

19. D. D. Palmer, D. K. Rehbein, J. F. Smith and O. Buck, J. Nondest. Eval. in press.
20. H. Tada, P. Paris and G. Irwin, "The Stress Analysis of Cracks Handbook," Del Research Corporation, St. Louis, MO, 1973.
21. R. Yeske and J. Weertman, Met Trans. *6A*, 1127 (1975).

Image Correlation on an IBM-PC for Characterization of Advanced Ceramics

J. J. Vaitekunas, R. A. Roberts, and W. A. Ellingson

Materials and Components Technology Division
Argonne National Laboratory
Argonne, Illinois 60439

ABSTRACT

Digital images generated from data obtained with ultrasonic scans, radiography, x- and gamma-ray computerized tomography, and nuclear magnetic resonance (also known as magnetic resonance imaging) often allow rapid visual interpretation of the data. Methods to compare the data generated by different techniques would be very useful. An inexpensive, high-resolution microcomputer-based system has been developed at Argonne National Laboratory to provide a means of correlating information in images generated by different NDE methods. This paper describes the imaging system and some of the techniques used to aid in correlating features in side-by-side images generated by the different methods.

INTRODUCTION

The reliable use of ceramics as structural components requires effective failure prediction and thus effective nondestructive materials characterization and flaw de-

---

To be published in the Proceedings of the Conference on Nondestructive Evaluation for Aerospace Requirements, sponsored by the University of Alabama (Huntsville) in cooperation with the Aerospace Committee of ASNT and the Huntsville Sections of ASM and AIAA, August 26–27, 1987, Huntsville, AL.

tection capabilities. Efforts to develop nondestructive methods to detect variations in material properties and other defects may be highly cost effective. Such methods are under development at Argonne National Laboratory.[1] However, each NDE modality is sensitive primarily to one parameter; e.g., x-ray computerized tomography is sensitive to density variations, nuclear magnetic resonance to chemical variations, and ultrasound to mechanical properties. Thus, it is of interest to compare features detected by different NDE modalities.

This paper describes a cost effective and useful way of displaying digital data acquired from various NDE interrogation methods. Also described is a software program that is under continuing development to aid in the correlation of the various NDE techniques. A unique approach to aid in the interpretation and correlation of some of the techniques is described, as well as a method to implement some of the ideas on any imaging system.

HARDWARE OF THE SYSTEM

Figure 1 shows a block diagram of the system assembled for the imaging correlation studies. The imaging capability was built around an existing IBM-PC. Image files are approximately 0.25 megabyte (MB) long, but they are created from data files that may be over 8 MB long, so storage capability of 10 MB is a minimum requirement. An Iomega Bernoulli Box was added to provide the necessary image storage and retrieval capability. The Bernoulli Box contains removable disks, each about the size of a notebook, with 10 MB of data storage per disk. It provides the input-output (I/O) space and speed of a hard disk with the storage and archiving convenience of a floppy disk. The slow I/O speed of the floppy disk also makes a hard disk more attractive because of the time saved on image retrieval and display.

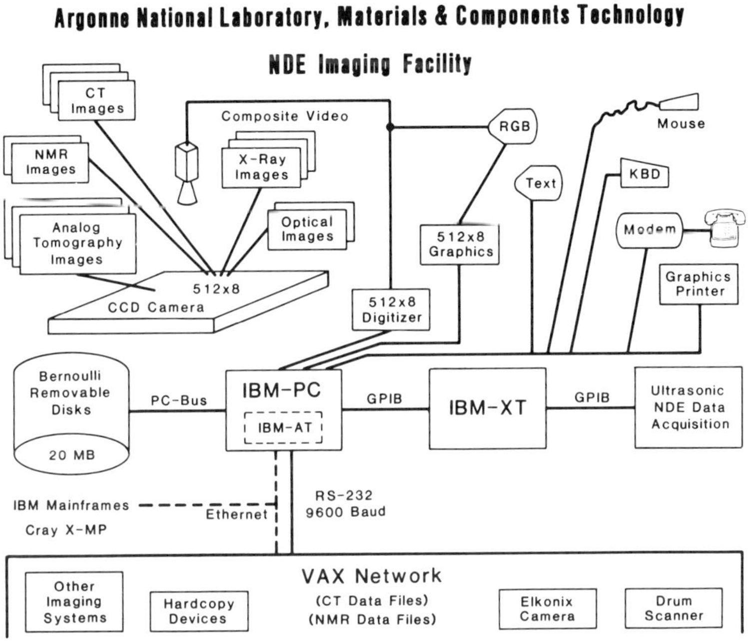

FIGURE. 1- Block Diagram of NDE Imaging Facility.

The imaging capability is based on a Hitachi high-resolution monitor, (100-MHz bandwidth) and an Omnicomp 1400 GDC graphics card. The Omnicomp card has a resolution of 2048 × 1024 with 8 bits of brightness resolution (shades of gray), and 2 bits for overlays. The camera is a Sony CCD (Charge-Coupled Device) video camera with a resolution of 512 horizontal by 483 vertical picture elements in the CCD array. The output of the camera goes to both the monitor and the input of a Chorus Data Systems digitizing card. Since the standard RS-170 video signal is used, any image acquired with this camera will have a non-square pixel shape. The monitor is used to focus and align an image from one of the NDE modalities: the digitizing card then digitizes the image with a resolution of 512 × 483 × 8. Digital data, such as ultrasonic scans, can be imaged directly without going through the digitizing card. An expansion chassis was needed to allow the addition of all the necessary cards to the IBM-PC.

SOFTWARE OF THE SYSTEM

Most of the imaging software was written in Fortran. Fortran was chosen because of the large quantity of scientifically oriented support software that is commercially available. However, one problem with current PC Fortran compilers is their inability to manipulate single-byte unsigned integers. Unsigned byte integers are the standard format used when placing or retrieving a line of image data on the graphics card. Any program that does mathematical manipulations on an image requires that the image line data be transformed into a format suitable for math operations. A very efficient "trick" for obtaining fast data transformation can be played on Fortran compilers (Microsoft Fortran, for example) that allow the conversion of integer arrays to character arrays via EQUIVALENCE statements. With this trick, a 512-byte row of unsigned single-byte integers can be transformed into an array of 2-byte signed integers, as shown in the following example.

A line of 512 pixels is put into a 256-dimensioned 2-byte integer array. The high byte and low byte of each 2-byte integer must be separated and placed into a 512-dimensioned array for any mathematical pixel manipulation. By equivalencing screen integer arrays (2 bytes by 256) to character arrays (1 byte by 512) and then equivalencing 2-byte integer arrays (dimensioned 512) to character arrays (dimensioned 2 by 512), transferring screen data into a Fortran-usable math format can be done as a simple one-line character manipulation, which can greatly reduce the execution time of some image-processing programs. An example of Fortran code for this process is shown in fig. 2.

A unique data-file-to-image mapping routine was also developed to allow the entire ultrasonic waveform associated with a pixel on an image to be displayed or manipulated for further analysis. Instead of just the transit time or amplitude of an ultrasonic x-y scan, the entire waveform at each point is saved in a disk file. Amplitude, transit time, or other criteria are extracted from the waveform data and imaged. The image is visually examined to find points of interest, and then the entire waveform of any point can be recalled to help define the origin of a piece of pixel data. Localized groups of waveforms can then be further processed with more advanced techniques. This approach allows only areas of interest to be worked on instead of an entire image, and makes advanced signal analysis on a PC feasible.

Figure 3 shows an ultrasonic through-transmission image created from an 8-MB data file (256 × 256 pixels with a 128-point waveforem corresponding to each pixel). Each pixel in the image is mapped to its corresponding waveform in the data file. Moving a cursor to the pixel of interest and pressing a button on the mouse automatically displays the corresponding waveform. Three waveforms corresponding to three positions on the image of fig. 3 are displayed in figs. 4–6. The three positions are,

```
C     DATA TRANSFORMATION TRICK
      INTEGER*2 INTLINE(256),UNCOMPRESS(512)
      CHARACTER*1 CHARLINE(512),CHARUNCOMP(2,512)
      EQUIVALENCE (INTLINE,CHARLINE)
      EQUIVALENCE (UNCOMPRESS,CHARUNCOMP)
C     READ IN A LINE OF SCREEN DATA TO INTLINE
      DO 1 I=1,512
    1 CHARUNCOMP(1,I)=CHARLINE(I)
C     UNCOMPRESS NOW CONTAINS THE DATA IN USABLE FORMAT
```

FIGURE. 2- FORTRAN Code for Data Transformation.

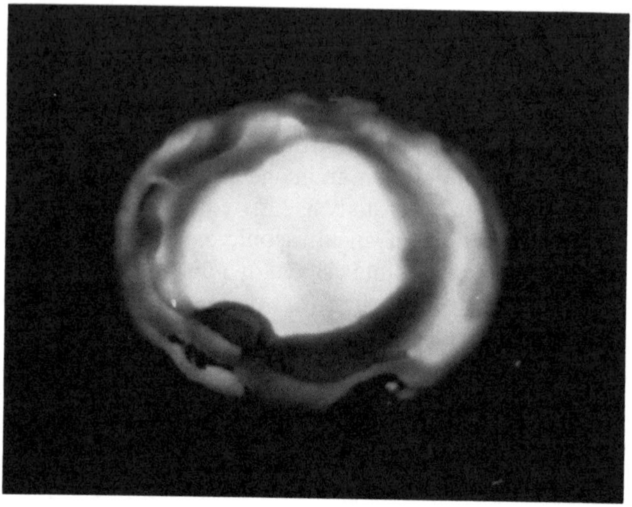

FIGURE 3- Ultrasonic NDE Image of Green-State Ceramic Sample.

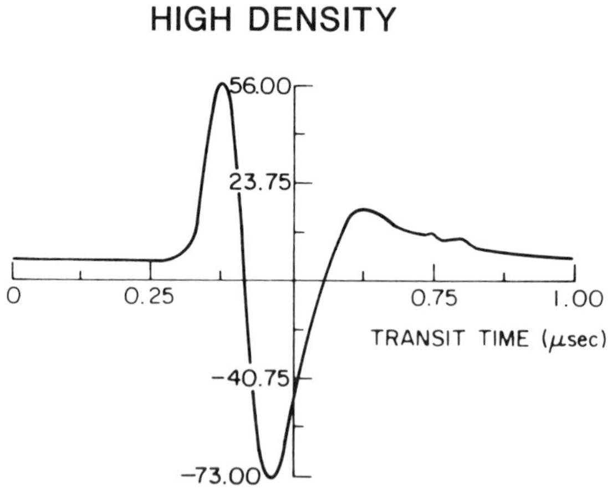

FIGURE 4- Ultrasonic Waveform from a Dark Region of Fig. 3.

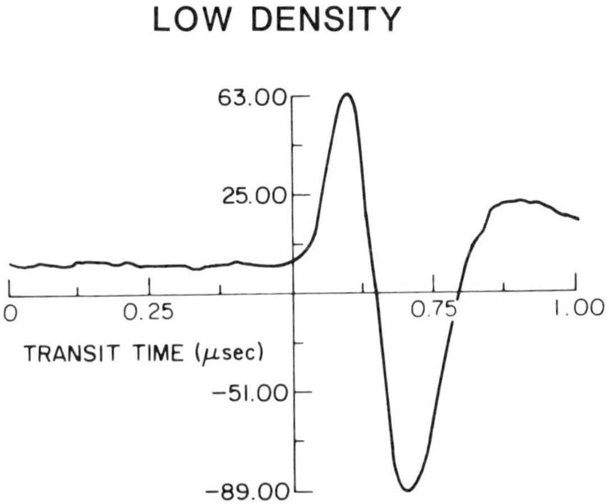

FIGURE 5- Ultrasonic Waveform from a Light Region of Fig. 3.

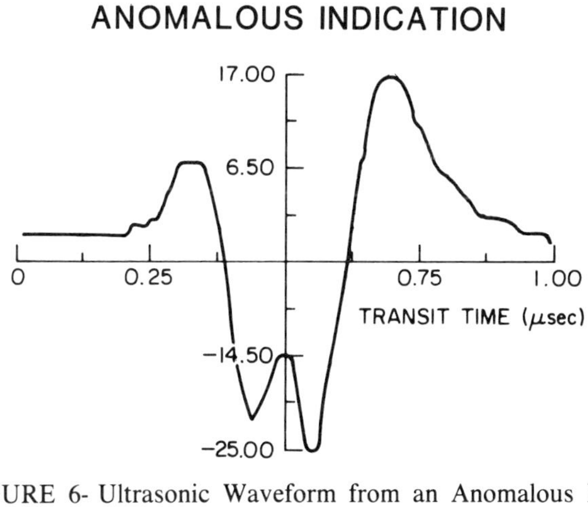

FIGURE 6- Ultrasonic Waveform from an Anomalous Inidcation Seen as a Light Region in Fig. 3.

respectively, a high-density area; a low-density area; and an anomalous indication which shows up as a low density area.[2] The basis for each pixel value could not have been determined conclusively without seeing the entire waveform. This method of retrieving the waveforms has allowed many insights into what is happening acoustically to create these images.

CORRELATION METHODS

Several problems are encountered in correlating features of images obtained from several NDE modalities. Some procedures produce non-square pixels, which are difficult to overlay on top of images with square pixels. Any digitizer that receives its input from an RS-170 (standard video format) camera will produce an image with rectangular pixels, having an aspect ratio of about 4 to 3. But most translation stages are incremented the same amount in any direction, providing square image data. A sample being imaged may be rotated differently in one mode as compared to another. Different magnifications, non-uniform magnifications along different axis, and mirror images are all encountered.

A simple and effective alternative to image overlay is the use of two cursors moving in relative coordinate systems on the graphics display. Figure 7 shows a split-screen display, which provides two images of the same sample interrogated with two different NDE modalities. The left half is a plot of a maximum-amplitude through-transmission ultrasonic scan. The right side is a digitized contact x-ray image. The x-ray image is rotated, magnified, and reversed relative to the ultrasonic scan. Also, magnification is different in the horizontal direction than in the vertical direction. Determining the relationships between the two images by mental comparison is not

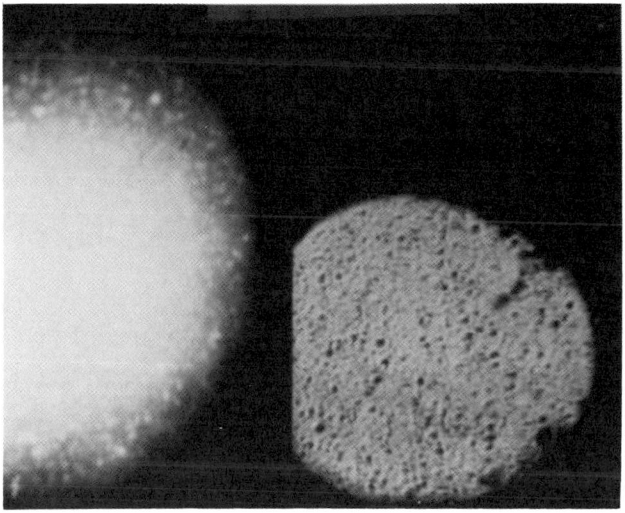

FIGURE 7- Split-Screen Image of a Densified Ceramic Sample. (Left) Ultrasonic image; (right) contact X-ray image.

easy. One way to overcome this problem is to display two cursors on the screen at the same time, one on each image, always in analagous positions relative to the coordinated systems of their respective images.

Normally, when two side-by-side images are to be compared on a graphics monitor, two coordinate systems are set up and the relationships between the two systems are defined to calculate the position of a point in one system relative to the other system. Instead of thinking of translation, rotation, magnification, and mirroring as separate entities, the relationship between the two coordinate systems is treated as two related functions existing on the screen's coordinate system (see fig. 8). The relationship can be expressed as in fig. 9 by expanding this relation-

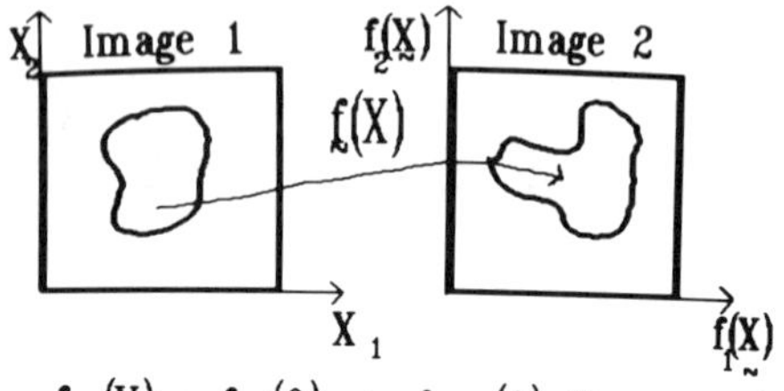

$$f_i(\underset{\sim}{X}) \approx f_i(0) + f_{i,j}(0)\,X_j + \dots \qquad i=1,2 \quad j=1,2$$

$$f_1 \approx f_1(0) + f_{1,1}(0)x_1 + f_{1,2}(0)x_2 + \dots$$

$$f_2 \approx f_2(0) + f_{2,1}(0)x_1 + f_{2,2}(0)x_2 + \dots$$

FIGURE 8- Relationship between Split-Screen Images.

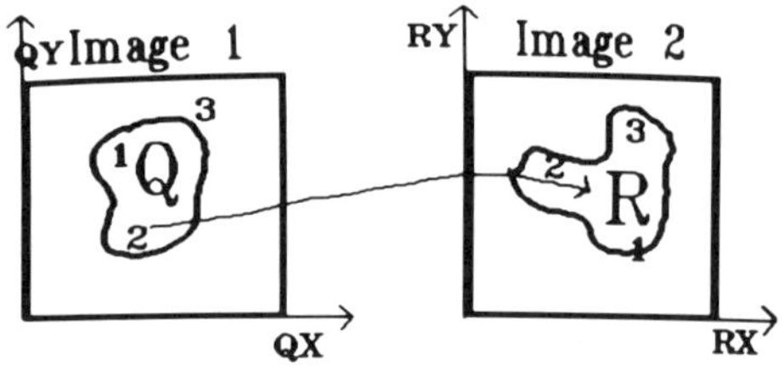

$$\begin{vmatrix} Q1x & Q1y & 0 & 0 & 1 & 0 \\ 0 & 0 & Q1x & Q1y & 0 & 1 \\ Q2x & Q2y & 0 & 0 & 1 & 0 \\ 0 & 0 & Q2x & Q2y & 0 & 1 \\ Q3x & Q3y & 0 & 0 & 1 & 0 \\ 0 & 0 & Q3x & Q3y & 0 & 1 \end{vmatrix} \begin{vmatrix} a11 \\ a12 \\ a21 \\ a22 \\ c1 \\ c2 \end{vmatrix} = \begin{vmatrix} R1x \\ R1y \\ R2x \\ R2y \\ R3x \\ R3y \end{vmatrix}$$

FIGURE 9- Computer Implementation of Truncated Taylor Series.

ship in a Taylor's series and truncating after the first order. It is apparent mathematically that by finding three point pairs between the two images and inverting the matrix, one can derive the vector containing the a and c coefficients that define the coordinates of the second image relative to the first. The transformation would have the form XNEW=A11*XOLD+A12*YOLD+C1 in Fortran.

Figure 10 shows the same image as fig. 7, with the addition of three labeled point pairs and a pair of markers (arrows) corresponding to the positions of the two cursors. When the mouse is used to move the cursor on one image, the algorithm calculates the position of the second cursor so quickly that the time delay before motion of the second cursor is not discernible to the eye. Any images

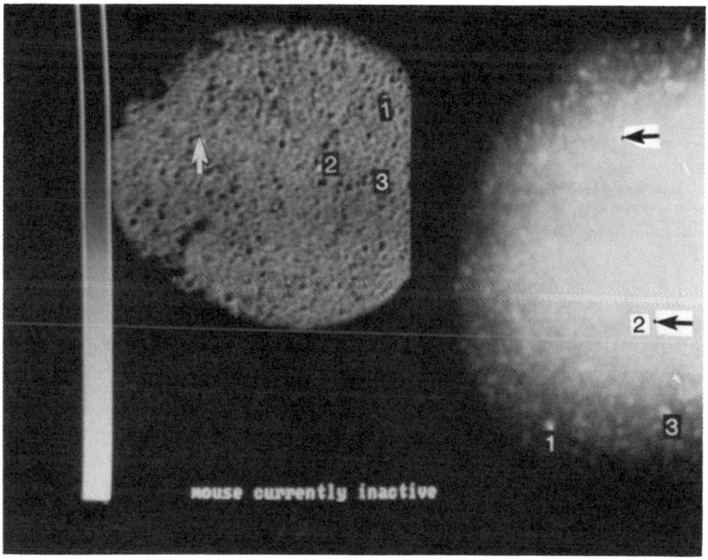

FIGURE 10- Correlation of Split-Screen Image.

displayed together can be used if three corresponding points can be found.

This dual cursor routine was used to correlate NDE data with fractography results on modulus-of-rupture (MOR) bars.[3] Silicon nitride (Si_3N_4) powder was hot pressed into disks, and these disks were cut into bars approximately 3 × 3 × 35 mm in size. The exact lengths were measured with calipers, and NDE methods were used to interrogate the bars. They were then subjected to four-point bending to failure. Fractography was done to find the failure-initiating flaw; during this process, pieces of the bar were destroyed. The calipers were then reset to the initial length of the bar, the remaining pieces of the bar were placed in them, and an image of the bar

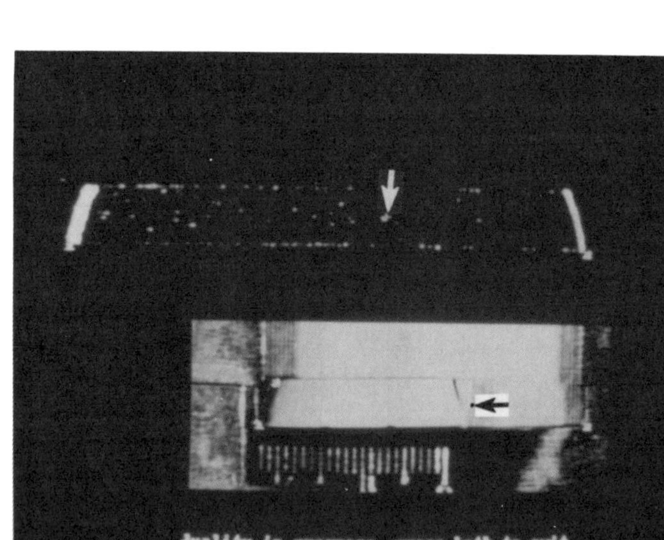

FIGURE 11- Correlation of Ultrasonic Backscatter Image with Fractography of a Silicon Nitride MOR Bar.

in the calipers was acquired by digitization. Although pieces of the bar had been lost, the calipers provided a reference for the three points required by the dual-cursor routine. The image of the broken bar was then displayed on the screen concurrently with the image obtained from one of the NDE modalities. The dual-cursor routine was implemented, so that when the cursor on the image of the fractured bar was moved to the position of the failure-initiating flaw, the other cursor simultaneously moved to the position of that flaw on the NDE image.

The screen display in fig. 11 shows a fractured bar in the calipers at the bottom, and an ultrasonic backscatter microscopy NDE image at the top.[4] The two bottom corners and the upper left corner of the bar were chosen as reference points. The bottom cursor was is placed over the failure-initiating flaw; the top cursor is seen directly over a spot in the NDE image that corresponds to the flaw before fracture.

SUMMARY

A state-of-the-art high-resolution imaging system has been developed at Argonne to aid in the correlation of features determined by different NDE modalities. The essentials of such a system can be reproduced by anyone with microcomputer. Through the development of appropriate software, it is possible to overcome the inherent limitations of microcomputer-based imaging systems relative to mini- and mainframe-based systems. For NDE applications where speed is not of primary concern, a microcomputer-based system offers a low-cost option.

Acknowledgments

Work supported by the U.S. Department of Energy, Office of Fossil Energy, Advanced Research and Technology Development Materials Program, under Contract W-31-109-ENG-38. We would like to thank Paul Raptis for his support and encouragement of this work.

References

1. W. A. Ellingson, R. A. Roberts, M. W. Vannier, J. L. Ackerman, B. D. Sawicka, S. Gronemeyer, and R. J. Kriz, "Recent Developments in Nondestructive Evaluation for Structural Ceramics," Argonne National Laboratory Report ANL/FE-87-2, May 1987
2. R. A. Roberts, "Ultrasonic NDE of Green-State Ceramics by Focused Through-Transmission," in Review of Progress in Quantitative Nondestructive Evaluation, Vol. 6B, D. O. Thompson and D. E. Chimenti, eds., Plenum Press, New York (1986), p. 1443.
3. J. P. Singh, R. A. Roberts, J. J. Vaitekunas, and W. A. Ellingson, "Failure Prediction of Hot-pressed Si3N4 Ceramics by NDE," in Proc. 32nd ASME Intl. Conf. on Gas Turbines, Anaheim, CA, May 31–June 4, 1987, American Society of Mechanical Engineers, New York (1987)), Paper 87/GT-7.
4. R. A. Roberts, J. P. Singh, and J. J. Vaitekunas, "NDE and Fracture Studies of Hot-pressed Si3N4," Ceramic Engineering and Science Proceedings, vol. 8, No. 7-8, (1987) pp 493–501

ZT: A Tomographic Backscatter Technique for Nondestructive Evaluation

Martin Annis and Paul Bjorkholm

American Science and Engineering, Inc., Cambridge, Massachusetts

ABSTRACT

A new x-ray imaging technique has been developed at AS&E which has unique capabilities. It is a tomographic, backscatter system called ZT. The technology is intrinsically one-sided, images thin slices of the object under inspection parallel to the front surface of the object, and is sensitive to the electron density of the material. The technique consists of a scanning beam of radiation appropriately chosen for the application, combined with a highly efficient collimator and x-ray detector system. The system is particularly sensitive to delaminations within the first ten or twenty centimeters of the material. Not only is the system more sensitive to the detection of delaminations than CT but because of the display, operators will perceive the delaminations more easily and at a smaller size. The technique is also particularly sensitive to water within structures. It can detect amounts of water that are as small as 0.1. cubic centimeter within composite materials. A laboratory system is described in detail, compared to CT, and applications of the technique to nondestructive evaluation will be given with quantitative results.

PRINCIPLE OF OPERATION

The technique requires two separate components, a source of scanning radiation and a collimated detector. The source produces a well defined pencil beam of radiation that can be moved within a fan shaped region. This is shown schematically in fig. 1. If the radiation back-scattered from the object were measured, the signal would be related to the scatter cross-section as integrated along the line of sight into the object times the flux at the given depth. Then, if a collimated detector is added, it is possible to differentially measure the scatter cross-section at a given depth in the object. This is shown in fig. 2. Here, the view is from the top. The collimator consists of multiple channels defined by x-ray attenuating material. In an apparatus we have constructed, each channel has parallel sides and is designed such that the projections of the sides intersect the plane defined by the scanning beam to define a slice thickness in the beam plane. The slice thickness and the distance of the intersection from

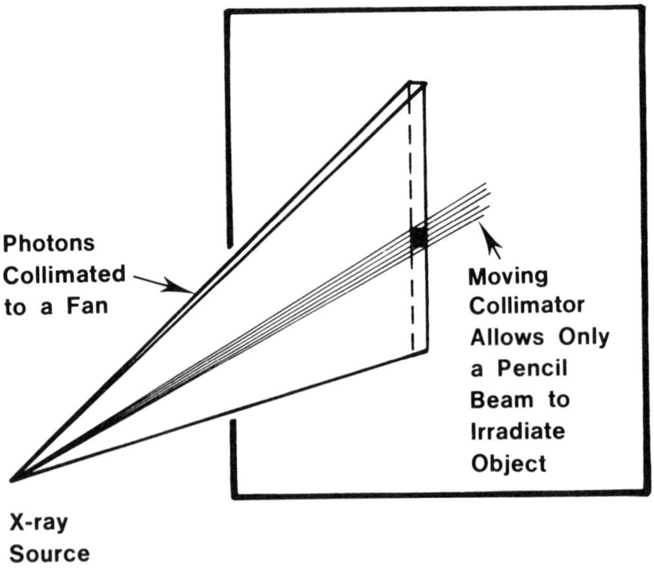

FIGURE 1- Scanning x-ray Beam Technique.

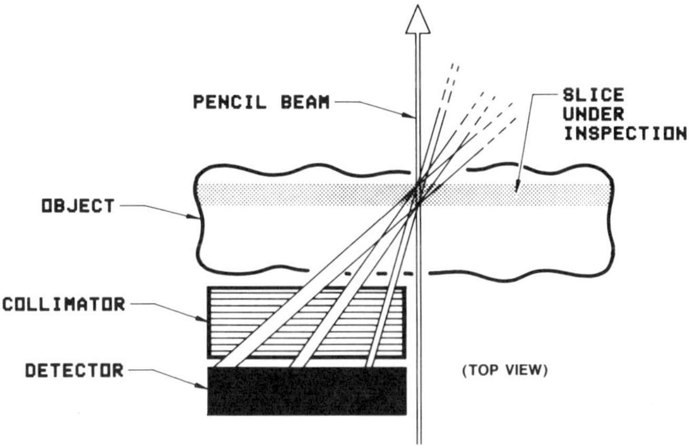

FIGURE 2- Schematic indication of how the linear collimator and pencil beam define a sub-surface slice.

the front of the collimator are design parameters. The detector can only see scatter originating from this volume of intersection. Since the illuminating radiation is in the form of a pencil beam, only a small section of the total volume is illuminated at one time. Therefore, the detector responds only to radiation scattered from that small sub-surface section of the object. As the beam is scanned, the total sub-surface volume is traced out. If the object is then moved in a direction perpendicular to the plane of the beam (or if the source, collimator and the detector are moved relative to a stationary object) at a speed slower than the x-ray scan speed, a slice parallel to the surface is traced out. If one samples the output of the detector repeatedly with a time constant small compared to the time it takes the beam to move over a region of interest, many samples of the scatter from the sub-surface plane may be obtained. These samples can be digitized, stored, and displayed as a visual image of the object. A schematic of a typical system is shown in fig. 3.

We have constructed a collimator which has the property of collecting scattered radiation with extremely high efficiency. This collimator is made up of flat planes. The collimator has the property that it may remain fixed as the flying spot moves parallel to the planes of the colli-

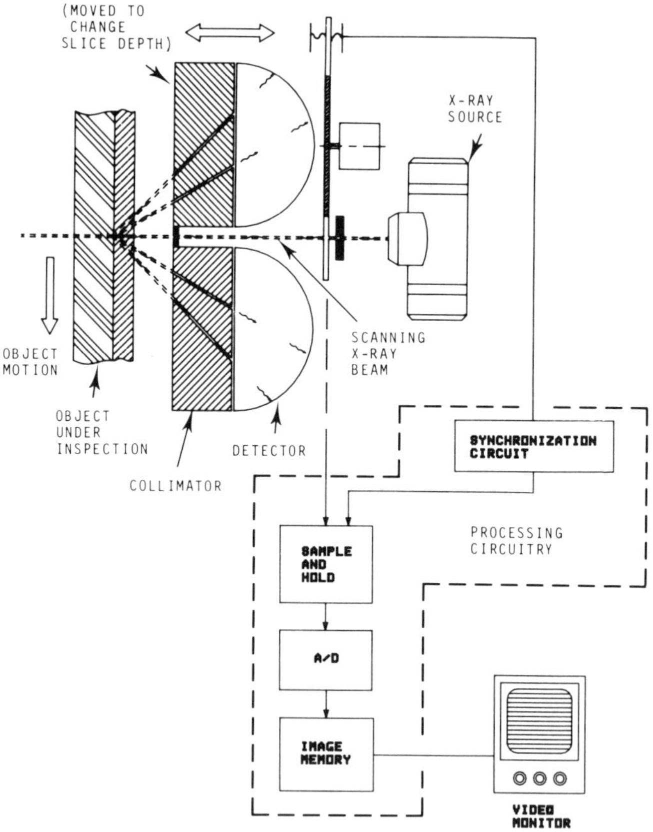

FIGURE 3- The ZT Imaging System Schematic.

mator. The principal disadvantage of this collimator is the fact that double scattering events originating from primary scattering interactions by the pencil beam not in the slice plane cause extraneous signals. Thinner slices reduce this effect. We have also designed a conical collimator which minimizes this effect.

THEORY

X-rays interact with matter via a number of different phenomena. These are principally the photo-electric effect, pair production, and Compton scattering. Some minor interactions will not be addressed here but can be found in the literature.

The photo-electric effect causes the removal of an x-ray photon from the beam with the energy converted into kinetic energy of an electron and, sometimes, lower energy fluorescent photons. The detailed energy conversion depends upon the atomic number of the material (Z). The fluorescent radiation from the slice is also detected and may be important for high Z materials.

X-rays at an energy much greater than 1 MeV can interact with matter by pair production. Here the photon energy is converted into a pair of particles consisting of an energetic electron and an energetic positron. The positron will eventually annihilate with an electron and yield a pair of 511 keV photons.

Compton scattering is caused by scattering of an x-ray photon by a nearly free electron. The energy of the scattered photon depends upon the energy of the incident photon and the angle of scatter. This effect can also be used to cast shadows in projection radiography but the scattered radiation causes a non-imaging background which is troublesome for most radiographic imaging systems. For example, in conventional radiography of human abdomens the flux of x-rays directly at the exit from the body is dominated by scattered radiation. Thus, successful diagnostic radiography of the human abdomen requires some form of scatter rejection.

As the energy of the incident photon increases, the average angle of scatter moves forward. However, even at very high energies some photons are scattered in the backwards direction. The differential scattering cross-section per unit solid angle is given by the well known Klein-Nishina formula.

This general review of interaction types between x-rays and matter is necessary because the type of imaging described here utilizes principally Compton scattered photons. For materials of high atomic number, the photoelectric effect will produce fluorescent K-shell photon emissions which occur with a high probability and which may also be used for ZT imaging. For the higher energies and/or the lower range of atomic number (Z), Compton scattering dominates. Figure 4 shows the region where the photo electric effect, Compton scattering, and pair production predominate.[1] For most materials, Compton scattering is the dominant process from 100 keV to 10 MeV. For lower Z materials, the lower energy limit is further depressed. This range of energies and Z values includes most of the region of interest for radiography, including medical applications. However, there are particular applications where the higher Z materials can be most effectively imaged by their fluorescence.

We will consider a typical application of the ZT technique with a convenient geometry which we have employed (see fig. 5). In this geometry which is relatively simple to calculate approximately, the pencil beam of x-rays is incident normal to the surface of the object. Although 180° backscattering offers the lowest differential

FIGURE 5- ZT system geometry which produces near-optimum results.

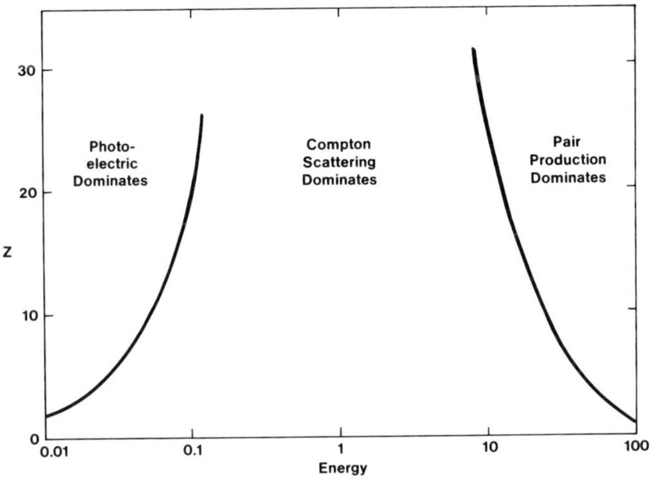

FIGURE 4- The three major x-ray interaction processes vary in probability with both atomic number and energy. The two heavy lines indicate the locus where the indicated process cross sections are equal. Note for low Z materials, Compton scattering dominates over a wide range of energy.

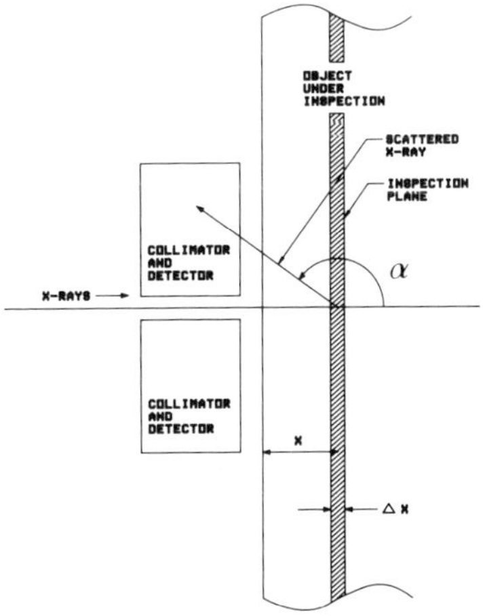

cross-section, and the scattered photon has the lowest energy at 180°, our calculations indicate this is an optimum geometry at all photon energies because (1) it allows the source to be closer to the object giving a greater flux, and (2) the path length of x-rays into the object is minimized. In this geometry, the back scatter detectors are placed on either side of the pencil beam plane as shown in fig. 2. In these calculations we assume that the input beam of photons is monoenergetic. For a polychromatic spectrum the equations below may be integrated over the input energy range. The scattered flux per unit area in the slice I_s (photons/cm^2) is given by:

$$I_s = I_0 e^{-\mu_0 x} \int_\Omega d\omega \frac{d\sigma}{d\omega} \frac{NZ}{A} \Delta x \; e^{-\mu' x/\cos(\pi - \alpha)} \; T(\alpha) \; \epsilon_{det} \qquad (1)$$

where

I_0 = incident flux (photons/cm^2)

μ_0 = total attenuation coefficient of incident x-ray (cm^2/g)

x = depth of slice (g/cm^2)

Ω = the solid angle subtended by the collimator at a typical point in the ZT slice. For ease of calculation, we have assumed that the excursion of the pencil beam across the slice is very much smaller than the height of the collimator. Thus Ω is assumed to be constant.

α = angle between incident x-ray and scattered x-ray

$\frac{d\sigma(\alpha)}{d\omega}$ = Klein-Nishina differential scattering cross-section, which depends upon the incident photon energy and the angle which in turn determines the scattered photon energy (cm^2/electron).

N = Avogadro number = 6×10^{23} atoms/molecular weight
Z = atomic number
A = molecular weight (g)
Δx = thickness of slice (g/cm^2)
$\mu'(x)$ = the total attenuation coefficient of scattered photons (cm^2/g), which depends on α.
$T(\alpha)$ = the transmission of the collimator
ϵ_{det} = efficiency of detection of the scattered photons (assumed to be ~0.3)

The factor $e^{-\mu_0 x}$ in Eq. (1) is the attenuation of the incident beam prior to reaching the ZT slice depth. Since this factor multiplies the factor which expresses the scattering from the slice, the resultant image density will be modulated by the over-lying material. Thus, the ZT image does not give an absolute quantitative measure of the density within the slice as does a CT image. However, since the factor is multiplicative, geometrical shapes are easily understood and diagnosed.

The factor $e^{-\mu'(\alpha)x/\cos(\pi-\alpha)}$ which appears under the integral is due to the attenuation of the scattered photons exiting the object. This involves an integration over the total solid angle and averages over large areas thereby reducing local overlaying density variations. In many industrial or medical applications the material overlying the ZT slice is very close to a constant thickness and composition so that quantitative density measurements may be achieved. Table 1 tabulates some useful parameters for several energies and angles of scattering. Note that the energy of the backscattered photons are kinematically limited. At 180° the scattered photon energy asymptotically approaches 255.5 keV. For the collimator and detector we have constructed $\Omega \approx 3$ steradian. Then Eq.

(1) becomes (approximately):

$$\frac{I_s}{I_0} = 4.15 \times 10^{22} \frac{d\upsilon}{d\omega} e^{-\mu_0 x} e^{-\mu' x/\cos(\pi - \alpha)} \qquad (2)$$

where

$\alpha \approx 150°$
$\epsilon_{det} \approx 0.3$
$T(\alpha) \approx 0.6$
$Z/A = 0.48$. Z/A is approximately constant aver the periodic table and has been chosen to be that of A1.
$\Delta x = 1$ g/cm^2 which is equivalent to 1.3 mm of F_e, 3.7 mm of A or 10 mm of plastic. The result may be easily scaled to other slice thicknesses.

Figure 6 shows an evaluation of Eq. (2) for various materials where the ordinate N_s, the number of scattered photons per cm^2 per milli-Roentgen of incident radiation is plotted as a function of the energy of the incident photon, E_0 for various amounts of overlaying material. The general features can be easily understood in terms of the physics involved. The monotonic decrease in the number of scattered photons at high energies reflects the decrease as a function of energy of the Compton cross section. The rapid decrease at low energies that depends on the Z of

TABLE 1-

Photon energy incident	Photon energy scattered at 135°	180°
100 keV	75 keV	72 keV
1000	230	204
10000	290	249

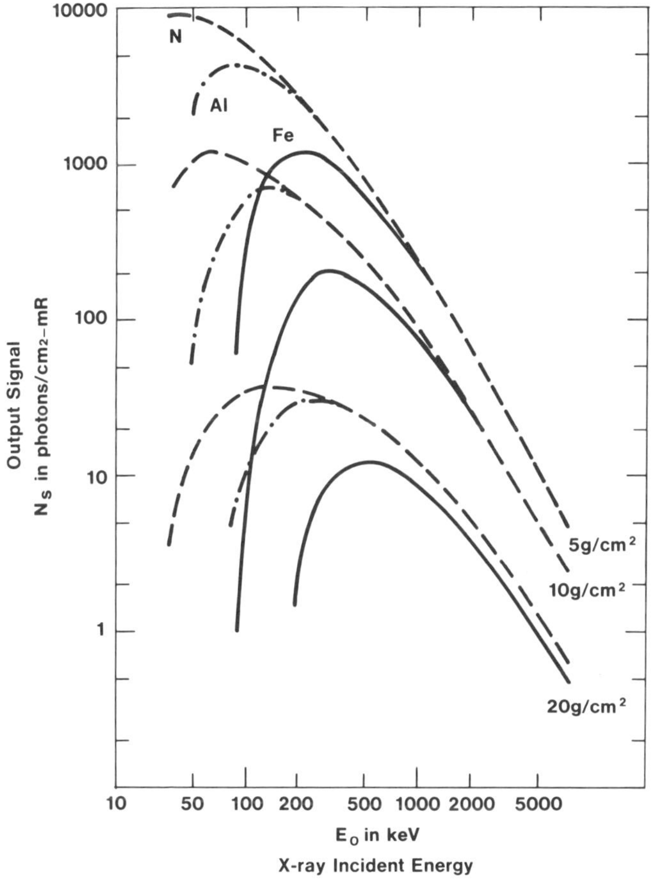

FIGURE 6- Number of scattered photons detected, N_s, per cm^2 and per milli-Roentgen of incident radiation as a function of the energy of the incident photon, E_0 for various materials and depths of slice.

the material is caused by the increase in the photo-electric cross-section as energy decreases.

Suppose, for example, one wishes to penetrate 5 cm of soft tissue (approximately 5 g/cm^2 of nitrogen) and achieve a 1 cm ZT slice. From fig. 6, it is clear than an x-ray peak energy of about 100 keV would be suitable and would result in a scattered flux of about 6000 photons/cm^2 $-$mR incident. In our laboratory we have achieved 10 mR incident over a field measuring about 12 cm $\times$ 12 cm in a period of 10 seconds with a pencil beam 1 mm $\times$ 1 mm. The corresponding spatial resolution for a flying spot system may be shown to be one line pair per millimeter.

Using an incident flux of 10 mR, there would be about 6×10^4 photons/cm^2, giving a one sigma standard deviation noise to signal of 0.4% per cm^2 or 4% per mm^2. The images we have achieved in our laboratory approximately affirm these estimates.

Consider first the problem of imaging a very thin object. Using conventional projection radiography, one would wish to utilize a low keV energy beam to improve contrast. This would be impossible if the thin object had a relatively thick structure behind it that had to be penetrated. The object would then have to be x-rayed with a high-energy beam, thus reducing contrast resolution, even at high x-ray energies. This is because *only* those photons which interact in the thin object are detected, resulting in a very large statistical signal-to-noise.

Figure 7 shows an example of a thin object. An automobile dashboard was given to us for evaluation. It consists of a metal base, $\frac{1}{4}''$ to $\frac{5}{8}''$ of low density foam, and a vinyl covering. The manufacturer was interested in detecting voids in the foam under the vinyl. Standard projection radiography could not demonstrate the void because the low density foam provided inadequate contrast. Backscatter radiography was also unable to demonstrate

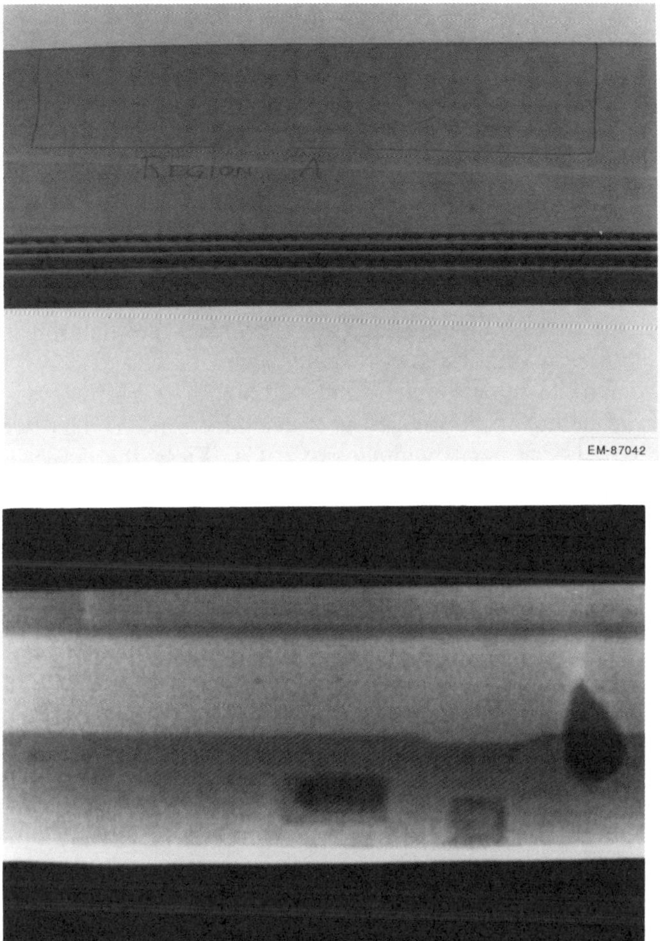

FIGURE 7- The top image is a photograph of a dashboard inspected by ZT as shown in the bottom image. The void in the foam is obvious in the ZT image. A projection radiograph does not show the void because of the very low radiographic density. A non-tomographic backscatter image does not show the void because of the overwhelming scatter from the metal back.

the void because of the very large scatter signal from the metal base. However, a ZT slice through the defect visualized the void with very high signal to noise. The increased signal to noise (as compared to standard backscatter imaging) is a result of the reduction of signal from scattering mass in front of and behind the ZT slice without any reduction in the signal from the void within the slice. This therefore represents a case where the material of interest is thin radiographically.

On the other hand, when the object is very thick and the defect to be imaged is a delamination, ZT has two very important advantages over CT. First, the defect is displayed *en face*. That is, for a 1 cm × 1 cm × 0.025 cm defect, the defect appears 1 cm × 1 cm in the ZT image. It is easily detected and its spatial extent is quickly and naturally seen and appreciated. While the thickness of the delamination is not immediately obvious, the intensity of the signal is proportional to the mass defect.

A CT display of the same defect would be a *line* of pixels 1 cm long. Since, for most CT systems, a 0.025 cm crack would be very much smaller than a pixel size, imaged crack width would be one pixel. Only one dimension would be displayed, the length of the object in the plane. The out of plane dimension could be determined only with multiple slices.

Figure 8 shows the displays one expects from ZT and CT of similar defects. Note that one appreciates different aspects of the delamination in the two images. Figure 8(a) shows a 0.040″ delamination in a large object utilizing a CT reconstruction using data taken with a 16 MeV x-ray spectrum. There is a clear and obvious change in the image at that location. Figure 8(b) shows a ZT image of a phantom which duplicates the densities and thicknesses of the object in (a) but is smaller and could be scanned on the laboratory ZT machine. Mass defects as thin as 0.005″ (127 microns) are easily seen and their dimensions

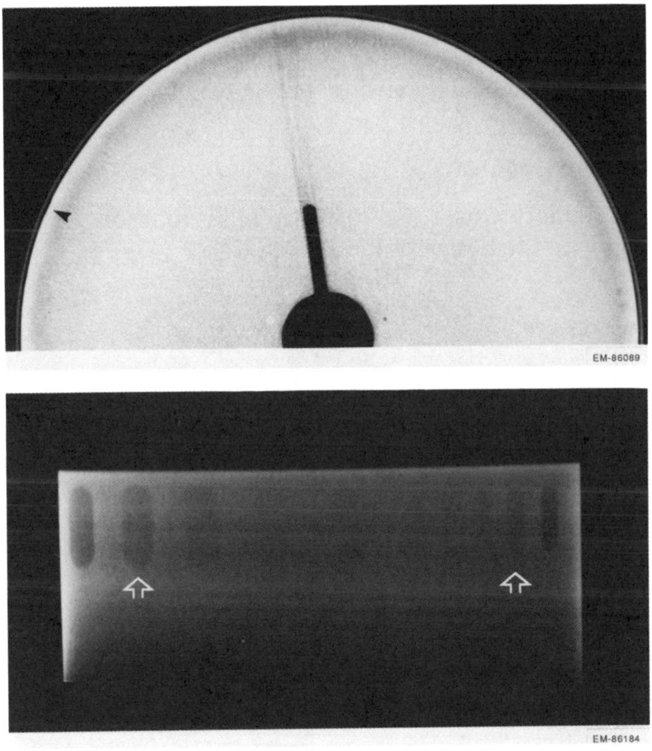

FIGURE 8- (a) A CT reconstruction of a phantom with a manufactured delamination of 0.040″ thick (black arrow). Note the defect appears as a narrow short line. (b) A ZT reconstruction of a series of manufactured delaminations of 0.080-, 0.040-, 0.020-, 0.010-, and 0.005-inch thick with two different frontal areas. The case thickness and densities are as in (a). The two white arrows note 0.040 inch delaminations. Note the ease of detection compared to CT. The 0.010-inch delaminations are easily seen in the original display. The delaminations decrease in the thickness from both edges to the center.

appreciated. These data were taken with a 110 kVp X-ray spectrum. This is an example of the power of this one sided perpendicular viewing x-ray modality.

Acknowledgments

The authors would like to thank Mike Johnson for his help with the design of the system.

References

1. Evans, R.D., 1955. The Atomic Nucleus (New York: McGraw-Hill).
2. McMaster, W.H., *et al.*, 1969. Compilation of X-Ray Cross Sections. (UCRL 50174 Section II, Rev. 1, Lawrence Livermore National Laboratory, University of California).
3. A.I.P. Handbook, 1972. New York: McGraw Hill.

A Review of Computer-Aided Interpretation Technology for the Evaluation of Radiographs of Aluminum Welds

J. F. Lloyd

Martin Marietta Manned Space Systems,
Advanced Quality Technology,
New Orleans, Louisiana

ABSTRACT

Industrial radiography is a well established, reliable means of providing nondestructive structural integrity information. The majority of industrial radiographs are interpreted by trained human eyes using transmitted light and various visual aids. Hundreds of miles of radiographic information are evaluated, documented and archived annually. In many instances, there are serious considerations in terms of interpreter fatigue, subjectivity and limited archival space. Quite often it is difficult to quickly retrieve radiographic information for further analysis or investigation. Methods of improving the quality and efficiency of the radiographic process are being explored, developed and incorporated whenever feasible. High resolution cameras, digital image processing, and mass digital data storage offer interesting possibilities for improving the industrial radiographic process. This paper will present a review of computer aided radiographic interpretation technology in terms of how it could be used to enhance the radiographic interpretation process in evaluating radiographs of aluminum welds.

INTRODUCTION

Radiographic film interpretation, evaluation and archiving are time consuming, laborious and expensive operations. There is a definite need to reduce operator fatigue, subjectivity of interpretation, storage space and inspection time. In response to that need Martin Marietta Manned Space Systems has been surveying components for integration into a computer aided radiographic film interpretation system for use in evaluation of aluminum welds. Components being evaluated are: high resolution cameras, digital image processing, optical disk storage and voice recognition.

SURVEY RESULTS

Cameras

The results of an inspection process are only as good as the weakest link in the system. Of primary importance is the basic detector, the camera, which in effect must match or exceed the capability of the human eye. There are several cameras available in both the Vidicon and

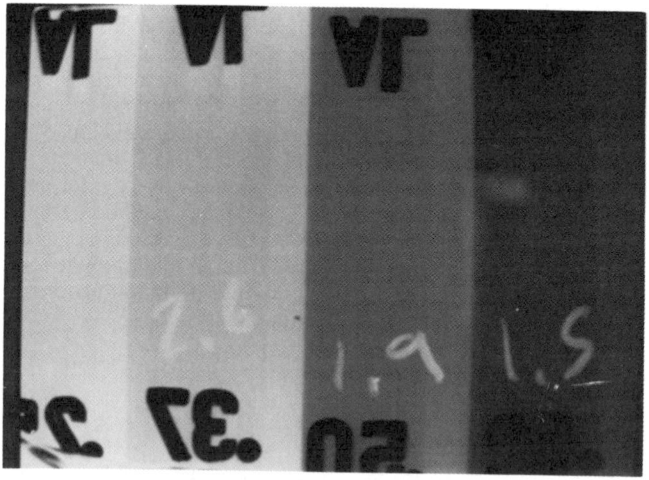

FIGURE 1- Detail and contrast sensitivity of penetrameter.

CCD categories which show acceptable image contrast and detail sensitivity detection capabilities. See fig. 1 for contrast and detail sensitivity in a radiograph of a penetrameter. Figure 2 shows Vidicons and a CCD camera. For this discussion, Vidicon cameras have been generalized as cameras employing electron tube type sensors. Vidicons represent a mature technology with improvements coming at a slower rate. They are generally larger. CCD cameras have been generalized as cameras employing solid state image sensors. They represent a rapidly advancing technology. They are generally smaller and more rugged. Each has its own advantages and disadvantages.

The Vidicon camera's horizontal resolution is determined by the video amplifier's bandwidth and the quality of construction of the vidicon itself. The vertical resolution is determined by the number of horizontal scan lines. The video signal of the Vidicon camera is relatively noise free and follows the imaged object in a continuous fashion, thus the sampling analog-to-digital converter (A/D) can readily pick up a representative digital image. The

FIGURE 2- Vidicon and CCD cameras.

vidicon is susceptible to some extent to external magnetic fields, blooming and lag, and may suffer permanent damage from burning an image into its photosensitive surface.

The CCD camera, on the other hand, is much more immune to image burning and blooming. Lag and magnetic fields are not problems. Its resolution is fixed by the geometry of the CCD chip array. The specification of this resolution and the actual number of photosites contributing to the picture sometimes appear in conflict. Care must be taken when interpreting this specification. The video signal of the CCD camera can contain a significant amount of clock noise in the time between actual pixel element signals. The sampling A/D must be synchronized to pixel rate and in the proper phase relationship or the clock noise will appear in the digitized image.

Figure 3 shows a temporary setup of camera and ra-

FIGURE 3- Camera and viewer.

diograph viewer. Proper selection of a camera and lens will depend on application requirements. Camera costs can range from $2,000 to $90,0000. CCD camera costs go to both price extremes while Vidicon prices fall into a smaller range near the low extreme. When balancing system requirements with equipment available, cost is a sizable factor to be considered. We found that a vidicon with 1600 TV lines and a CCD with 1024 × 512 elements were adequate for our requirements.

Monitors

Selection of a monitor depends on application requirements and must complement the rest of the system. A high resolution black and white monitor is sufficient for viewing original or enhanced radiographic images since there is no color information in the original radiograph. However, a high resolution RGB color monitor allows pseudo coloring which can be used to sharply distinguish very slight intensity changes. It should be noted that room illumination and brightness and contrast settings have a great effect on the apparent quality of the monitor.

Image Processing

Figure 4 shows an image processor used as part of a real time radiography system. Gray scale resolution and pixel resolution are basic considerations of an image processing system. Eight bits per pixel gray scale resolution is a reasonable choice for digitization and storage of images. While a higher number of bits appears better, the advantages would be in large part offset by increased storage requirements and increased noise pickup from the camera. There are dozens of image processing algorithms available to process and enhance data. (Figure 5 shows a frame averaged image of a weld flaw.) Which ones to

FIGURE 4- Advent II Image Processor.

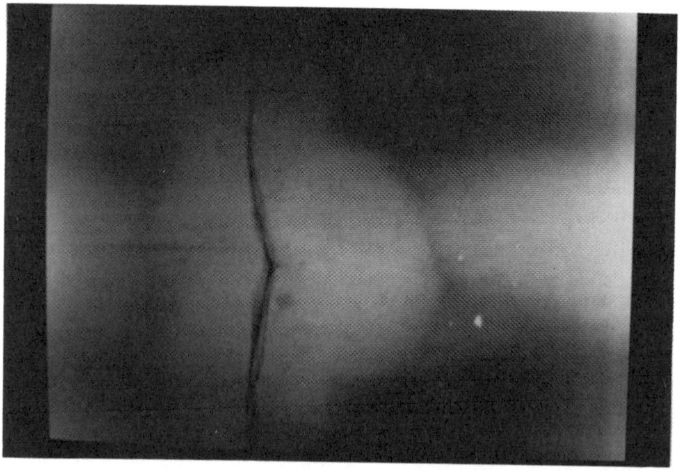

FIGURE 5- Weld flaw on monitor using frame averaging.

use depends on their particular effect on the information required. There are many usable filters which can improve the "seeability" of flaw indications in radiographs of welds. Examples are: Laplacian, Sobel, Roberts, median, homomorphic filters, histogram equalization and neighborhood averaging. Other utilities, such as zoom, scroll, pan, histograms, densitometric measurements, and micrometer capabilities, are also available. Additional information on typical image processing operations is given in Appendix 1.

Laser Disk Storage

Since the digitization and storage of the resulting data amounts to many gigabytes of digital information, the archival mechanism should be compact and readily accessible. Figure 6 shows a representative laser optical disk

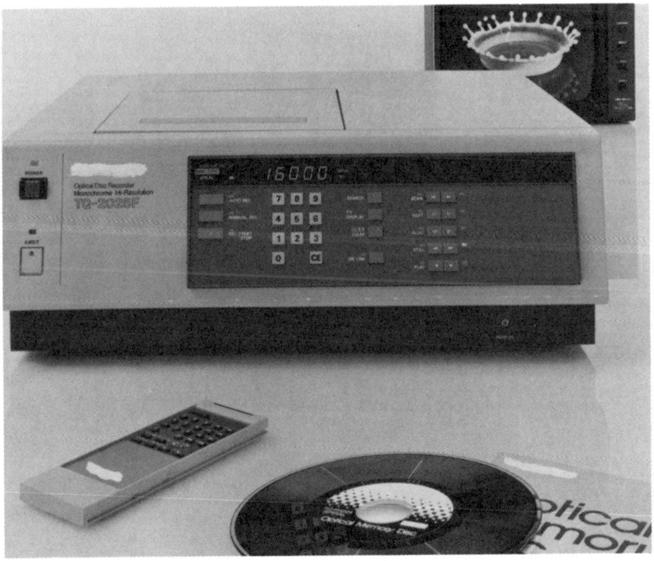

FIGURE 6- Laser Optical Disk Storage example.

storage subsystem. Laser optical disks have the capability of storing 2-gigabytes of digital information (1 gigabyte per side). Other smaller versions are also available. Even though 2-gigabytes capacity appears to be large, it is quite likely that various forms of data compression would be needed to further increase storage density depending on the quantity of film to be evaluated and stored. Laser disk storage also offers the advantage of .5 second recall time per frame of information. Film archival storage consumes a lot of space as shown by fig. 7. Additional time is consumed in retrieval by having to search serially through many feet of film to get to the section of interest.

Voice Recognition

Initial surveys indicate that voice recognition could be used advantageously to command a considerable portion of system operation. Figure 8 shows a representative

FIGURE 7- Film cannister archive.

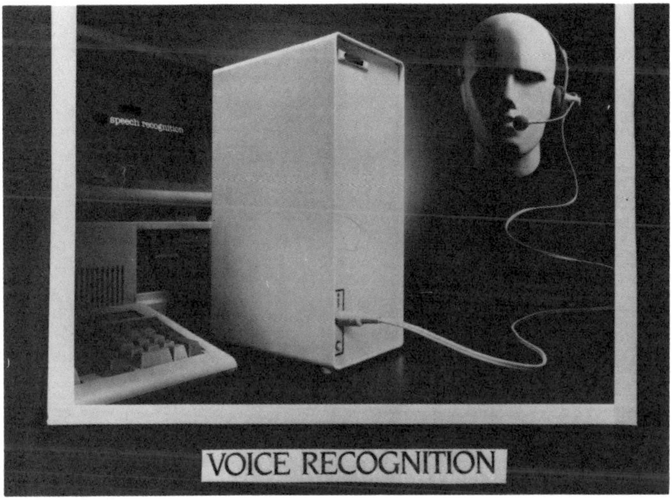

FIGURE 8- Voice recognition example.

voice recognition subsystem. Verbally entered evaluations and other pertinent information could be archived together with the actual digital information of the radiograph. More evaluation of real-world capabilities is necessary for a more definitive understanding of the voice recognition capabilities.

Computer and Peripherals

Figure 9 shows a laboratory setup of image processor, camera, film and viewer. There are several computers which would accomplish the requirements in a film interpretation system. The computer system arrangements for image processing can vary considerably. They can use general purpose micro, mini, or mainframe computers with image processing software or they can use special array and pipelined processor hardware under the or-

FIGURE 9- Image processor, camera, film and viewer.

chestration of general purpose computer. Others can be hardwired and hardcoded to perform a series of enhancement functions that are very fast but inflexible.

CONCLUSIONS

A computer aided radiographic interpretation system shows considerable promise as a viable tool to enhance the radiographic film interpretation process for aluminum welds. The major components for such a system have been identified. They are high resolution cameras and monitors, image enhancement hardware and software, laser disk storage and voice recognition subsystems. While many of these components and systems have been evaluated, additional evaluation and data gathering are necessary. Toward this end, evaluation of a turn-key system on production film is planned for the near future.

APPENDIX 1

Typical Image Processing Operations

The following are simplified explanations of some of the digital imaging processing terms:

Convolution. Convolution is used for spatial filtering and finding image features. The convolution operation replaces a pixel's value with the sum of that pixel's value and its neighbors values, each weighted by a factor. The weighting factors are called the convolution kernel. These kernels can take on various sizes, however the typical kernels square arrays such as 3 × 3, 5 × 5, 7 × 7, etc. It is not necessary for the arrays to be square but a square one can be reduced to a linear or rectangular array by making the weighting factors zero for the row(s) or column(s) which are not desired. The arrays are usually of odd dimensions so that there is not a shift of the image and also so that the effect of the kernel can be symetric if desired. Convolution is computationally expensive, therefore, larger kernels significantly slow down image processing. Convolution results in a border of garbage around the image edge where the kernel cannot be properly applied to the edge pixels. The larger the kernels are and the number of times the convolution is done results in larger borders.

Differentiation. Differentiation can be used to sharpen the image. Differentiation quantifies the rate of change of the intensities of the image. Edges usually have significant intensity changes and they can be enhanced through differentiation. In the continuous realm a gradient vector at any point is formed by taking the partial derivatives in x and y at that point. Since we are dealing with digital images there are a finite number of points. The magnitudes of the gradient vector at each of these points is approximated by sums of differences. These gradient magnitudes are generated by applying various difference operators to the digitized image at each point. These operators are small orthorgonal arrays. The common approximation of the gradient g(x,y) is derived from the image f(x,y) by taking the sum of the absolute values of the differences of the intensities between the current x and y position and the next position in the x direction (x + 1) and the next position in the y direction (y + 1). Mathematically,

$$g(x,y) = \text{abs}[f(x,y) - f(x+1,y)] + \text{abs}[f(x,y) - f(x,y+1)].$$

The difference operators in this case are:

$$\begin{matrix} 1 & -1 \\ 0 & 0 \end{matrix} \qquad \begin{matrix} 1 & 0 \\ -1 & 0 \end{matrix}$$

Robert's Filter. The Roberts Edge Filter or Roberts Gradient is a variation of the gradient as described above. Mathematically,

$$g(x,y) = \text{abs}[f(x,y) - f(x+1,y+1)] + \text{abs}[f(x+1,y) - f(x,y+1)]$$

The difference operators in the case are:

$$\begin{matrix} 1 & 0 \\ 0 & -1 \end{matrix} \quad \text{and} \quad \begin{matrix} 0 & 1 \\ -1 & 0 \end{matrix}$$

Prewitt and Sobel.

Prewitt operators:

−1	0	1		1	1	1
−1	0	1	and	0	0	0
−1	0	1		−1	−1	−1

Sobel operators:

−1	0	1		1	2	1
−2	0	2	and	0	0	0
−1	0	1		−1	−2	−1

Laplacian. The Laplacian is an edge detection operator. It is an approximation to the mathematical Laplacian, that is the second partial derivative of f(x,y) with respect to x and y.

The Laplacian has two disadvantages: (a) directional information is not available, and (b) it enhances noise. One version of the discrete Laplacian is given by:

$$L(x,y) = f(x,y) - (1/4)[f(x,y+1) + f(x,y-1) + f(x+1,y)0 + f(x-1,y)]$$

This edge detection operator in kernel for is:

0	−1	0
−1	4	−1
0	−1	0

Another Laplacian edge detection kernel is:

−1	−1	−1
−1	8	−1
−1	−1	−1

Laplacian edge enhancement kernels have additional weighting at the center which retains the original image but enhances the edges. Here are two of these kernels:

$$
\begin{matrix} 0 & -1 & 0 \\ -1 & 5 & -1 \\ 0 & -1 & 0 \end{matrix} \quad \text{and} \quad \begin{matrix} -1 & -1 & -1 \\ -1 & 9 & -1 \\ -1 & -1 & -1 \end{matrix}
$$

Homomorphic Filtering. An image can be approximated by an illumination-reflectance model. The illumination is responsible for the dynamic range and in general is characterized by slow spatial variations. On the other hand, contrast is a function of the reflectivity of objects and varies abruptly particularly at object boundaries. A homomorphic filter which will affect low and high spatial frequencies differently can effectively compress the dynamic range and enhance the contrast simultaneously if the low frequencies are attenuated and the high frequencies amplified. This process is mathematically involved.

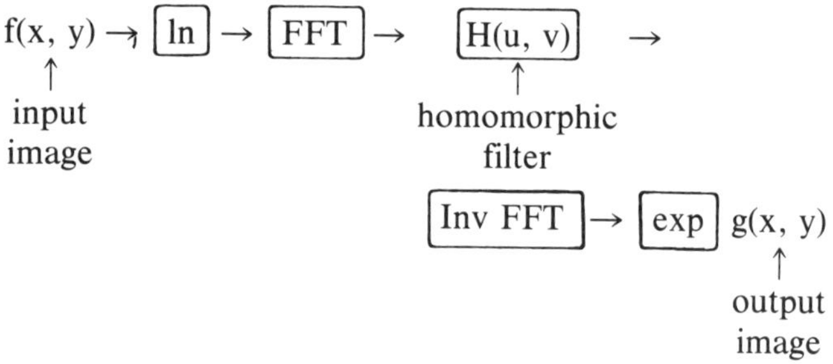

Time Averaging. Averaging multiple noisy images will remove a large portion of the uncorrelated noise. The improvement will depend on the number of images averaged. This is useful for image acquisition where the original image is time invariant, such as is the case in digitizing radiographic images.

Neighborhood Averaging. A pixel is replaced by a weighted average of its neighbors. This tends to remove noise; however, it also removes detail and causes edge blurring. Some averaging kernels are:

$$\frac{1}{5}\begin{matrix}0 & 1 & 0\\ 1 & 1 & 1\\ 0 & 1 & 0\end{matrix} \quad \text{and} \quad \frac{1}{9}\begin{matrix}1 & 1 & 1\\ 1 & 1 & 1\\ 1 & 1 & 1\end{matrix}$$

$$\frac{1}{10}\begin{matrix}1 & 1 & 1\\ 1 & 2 & 1\\ 1 & 1 & 1\end{matrix} \quad \text{and} \quad \frac{1}{16}\begin{matrix}1 & 2 & 1\\ 2 & 4 & 2\\ 1 & 2 & 1\end{matrix}$$

Median Filtering. A pixel is replaced by the middle value of its neighbors after they are numerically sorted.

*Optical Systems for Nondestructive Evaluation**

J. A. Gilbert, H. S. Johnson, B. R. Peters, and R. M. Ruggiero

Mechanical Engineering Department, University of Alabama in Huntsville, Huntsville, Alabama 35899

ABSTRACT

This paper describes some of the sophisticated optical systems currently under development at the University of Alabama in Huntsville which are being designed to inspect, detect and characterize structural integrity. The systems are based on various measurement techniques including holographic interferometry, high frequency moire interferometry, and thermal acoustophotonic nondestructive evaluation. The potential of these methods is enhanced by the introduction of fiber optic illumination and imaging components which permit remote operation and provide optical access to normally inaccessible locations. Computerized data acquisition and processing are incorporated into most of the systems to expedite analysis. These systems, either alone or in combination, can provide a wealth of information on the surface contour, surface displacement, and subsurface structure of a specimen.

INTRODUCTION

The development of future aerospace health monitoring and adaptive control systems will require major advances

* This manuscript was prepared for presentation at the Nondestructive Evaluation for Aerospace Requirements Conference held in the Huntsville Marriott, Huntsville, Alabama, August 26–27, 1987.

in sensor technology. An integrated approach of computer analysis and control with advanced sensors is key to producing a new generation of diagnostic systems which can detect anomalies, continuously monitor performance, or make critical decisions concerning safety and reliability. The authors are currently developing some of these sensors and diagnostic systems, most of which rely on fiber optic components.

The general scheme for making measurements using fiber optics is to have an illumination source (usually a laser or laser diode) to excite the optical fiber through a suitable coupler, an effective means of somehow modulating intensity, phase, polarization, or wavelength in response to the desired variable to be measured (force, pressure, temperature, strain, displacement, velocity, etc.) and a means of detecting and decoding the resulting modulated light wave signal. The transduction between the parameter to be measured and the light wave itself may be carried out either within the optical fiber, called intrinsic sensing, or outside the optical fiber, called extrinsic sensing. If the transduction is accomplished intrinsically, it may either be distributed along the optical fiber or localized at some especially prepared location or locations.

Fiber optics have many important advantages when incorporated into diagnostic systems. Since they are dielectric devices they are largely insensitive to electromagnetic interference. They may be readily adapted for use in harsh environments, and their dimensions and geometrical flexibility support compact, readily adaptable designs and facilitate access to remote or otherwise inaccessible locations. As will be described, both single and multimode flexible optical fiber elements, individually or in any of a variety of arrays, provide useful tools for many experimental mechanics applications ranging from direct excitation to remote sensing. Since it is not feasible to describe, in any sort of detail, all of the work being conducted by the authors in this summary, the review of a few illustrative examples primarily concerned with extrinsic sensing (displacement measurement and flaw detection) will have to suffice.

HOLOGRAPHIC/FIBER OPTIC SYSTEMS

The fundamental concept of holographic recording was demonstrated nearly forty years ago when Dennis Gabor suggested the process as a possible means of improving the resolving power of the electron microscope.[1] It was not until the early 1960's, when Leith and Upatnieks produced the first high quality holographic image using a strong coherent source,[2] that widespread attention was given to holography. The exactness of the holographic image made it invaluable for detecting faults by optical interference, and holographic interferometry has since become an important diagnostic tool in nondestructive testing.

To be useful for nondestructive testing in an aerospace environment, a holographic system should be flexible, versatile and self-contained. For example, optics should be easily movable to allow close matching of path lengths to satisfy coherence requirements and to gain access to a remote test surface. The ability to vary the object illumination and observation directions is desirable to permit measurement of different displacement components, while the ability to vary the angle of the reference beam provides a means of varying the spatial frequency content of the hologram itself. Mechanisms for fixing the illumination and observation positions with respect to the test part in order to measure relative movements on three-dimensional objects, and for adjusting the reference illumination position with respect to the hologram in order to compensate for rigid body motion, are also desirable. These conditions can be met by incorporating fiber optics into holographic systems.

Many workers have demonstrated the advantages of using flexible fiber optic illuminators for holography[3–8] and holo-interferometry.[9–12] Gilbert et al. and others have also demonstrated the superiority of singlemode fiber optic illuminators for a variety of applications including double-exposure,[12–17] time-average[18] and concomitant or real-time holo-interferometry[19] (both object illumination and reference beams). They also explored the use of flexible optical fiber imaging bundles to facilitate access to remote or obscured test sites (e.g. submerged)[20] while recognizing the need to provide a significant degree of mechanical immobility to multimode bundles to assure the needed modal (phase) stability. On the other hand, mixed mode holographic systems which do not entail the transmission of phase information via an image bundle, such as remotely generated holograms of ultra-low spatial frequency (ULF holograms), are quite stable and effective for both holography and holographic interferometry.[21,22] Dudderar and Gilbert[7] also demonstrated the use of fiber optic bundles for both illumination and image transmission in pulsed laser holography, and Bjelkhagen[23,24] has studied applications of optical fibers and optical fiber bundles to pulsed laser holo-interferometry. The use of such extremely short duration (20 ns) exposures somewhat reduces the need for mechanical stability and extends the range of measurement to highly dynamic, non-repetitive events.

Figure 1 illustrates a real-time holographic/fiber optic set up used to measure thermally induced deformations in a multilayer circuit board caused by power-up of a surface mounted leadless ceramic chip carrier. The chip carrier lid was bonded directly to a clear glass mounting plate and two mirrors were used to view the front and

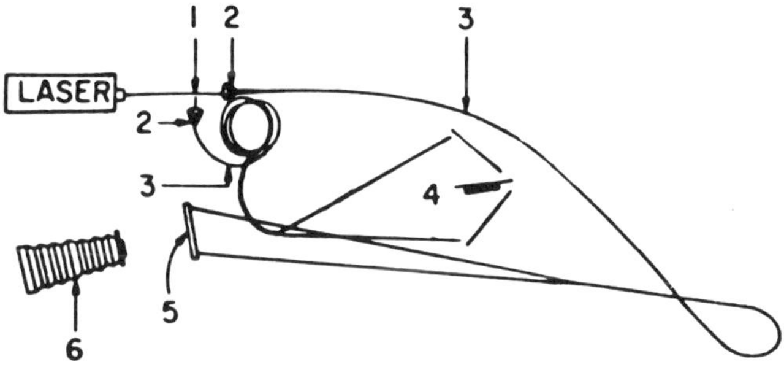

1-VARIABLE BEAM SPLICER 4-MODEL
2-FIBER LAUNCH OPTICS 5-HOLOGRAM
3-SINGLE MODE OPTICAL FIBER 6-CAMERA SYSTEM

FIGURE 1- Schematic of a real-time holo-interferometric fiber optic system for measuring out-of-plane displacements.

rear surfaces of the board simultaneously. A variable beam splitter (directional coupler) is used along with polarization preserving single mode optical fibers to generate the object and reference beams. Light exiting from the end of each fiber diverges to produce a coherent wavefront. Stability is enhanced, since each fiber acts as an environmentally less sensitive light guide (as opposed to propagation through air), and polarization can be carefully controlled by rotating the fiber end. The hologram itself is recorded using a holo-camera which produces and processes holograms suitable for real-time interferometry within seconds. Figure 2 shows a photograph of typical front and rear holo-interferometric fringe patterns.

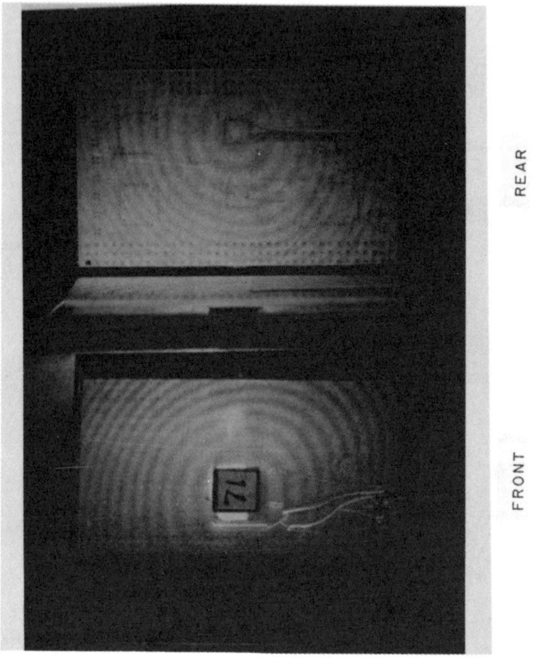

FIGURE 2- Holo-interferometric fringe patterns showing thermal deformation of the front and rear of a multilayer circuit board.

REAL-TIME MOIRE INTERFEROMETRY

High frequency moire interferometry is a highly sensitive full-field optical method useful for measurement of surface displacement.[25–27] The technique requires that a reflective high-frequency phase-type grating be transferred from a mold to the surface of the specimen. When the test surface is appropriately illuminated, the interference of the deposited grating with a reference grating having twice the spatial frequency yields a moire pattern that can be analyzed to measure a selected component of surface displacement.

In practice, aberrations and pre-load prevent the specimen grating from being perfect (i.e., composed of perfectly straight, uniformly spaced corrugations), and a few fringes may be observed in the initial field when it is interrogated by the reference grating. The introduction of a carrier fringe pattern prior to loading permits these initial "no-load" fringes to be optically subtracted from the load-induced displacement fringe pattern. Prior to the development of the real-time interferometric moire system described below, most studies required that a double exposure technique be used to optically remove the initial fringes from the field of view.[28] However, holographic techniques can now be used to store the initial carrier pattern for real-time analysis in the most general case.[29] The main advantage of this approach over the earlier double exposure technique is that the investigator is able to continuously monitor and evaluate full-field displacement during the loading process.

Figure 3 shows an experimental set-up used to observe and record real-time moire interferograms. The light beam from a laser was split to form the object illumination and reference beams required for holographic recording. Each beam was launched into a separate single mode step-index fiber;[12] used to reduce stray illumination, improve flexibility, and simplify the optical system. The exit end of the object illumination fiber was placed at the focal point of a parabolic mirror, producing a wide collimated beam of coherent light. The parabolic mirror directed this object beam towards the specimen holder assembly where a flat mirror was positioned at 90 degrees to the model. The entire assembly was rotated such that the $+1$ diffraction order of beam A coincided with the -1 diffraction order of beam B. With this alignment, light wavefronts were diffracted in a direction normal to the surface

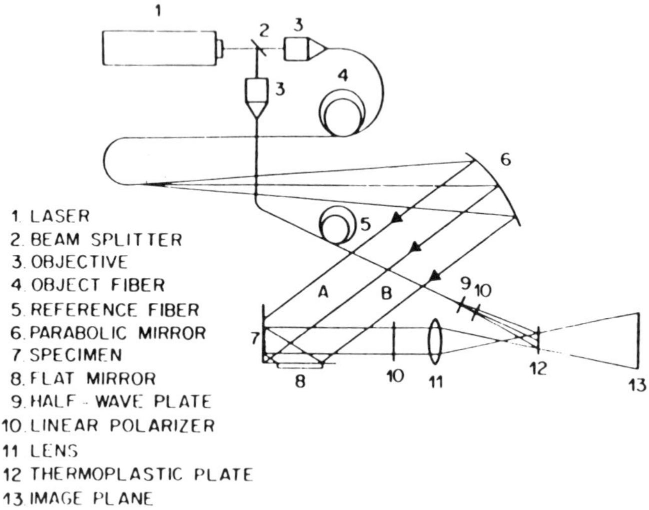

FIGURE 3- Schematic of a real-time moire interferometric system incorporating fiber optics.

and passed through a linear polarizer, an imaging lens, and a thermoplastic holographic recording device where they were stored holographically.

Light exiting from the reference fiber passes through a half-wave plate and polarizer before illuminating the thermoplastic holographic plate. The half-wave plate and polarizer combination provide variable light transmission in a preferred direction of polarization. The holographically recorded initial moire interferometric carrier is reconstructed using a horizontally polarized reference beam. This pattern is superimposed on the deformed carrier that is diffracted from the specimen grating in its loaded condition. In this case, the vertically polarized object beam and the horizontally polarized reference beam produce no interference and, consequently, no real-time holographic interference fringes. However, this process allows real-time interferometric moire fringes, such as those shown in fig. 4, to be photographed in the image plane. The pattern is indicative of displacement measured in a direction parallel to the longitudinal axis of the beam.

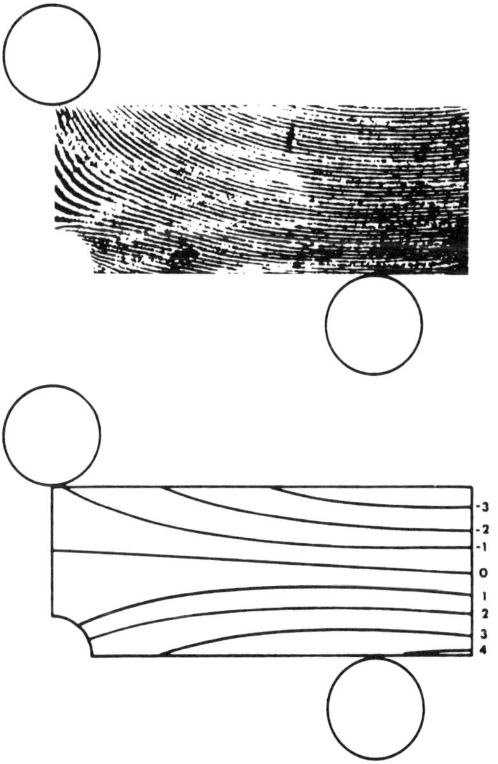

FIGURE 4- Real-time interferometric fringe pattern indicative of in-plane displacements, parallel to the longitudinal axis of the beam.

DIGITAL IMAGE PROCESSING AND COMPUTER ANALYSIS

One of the major factors inhibiting full exploitation of the holo-interferometric and moire interferometric techniques described above is the difficulty of getting numerical results out of the interferograms. For example, the acquisition of quantitative results from holo-interferometric patterns like those shown in fig. 2 often involves manual fringe order number assignment and hand entry of the data into a computer. Not only is this approach tedious and time consuming, but it is frequently subject to significant errors. The interferograms do not contain sufficient information to determine the direction or sign fo the surface displacement. Such inherent ambiguities must be reliably resoved if these data are to be used to accurately evaluate thermally induced warping and variations in the thickness change across the multilayer circuit board.

The automated quantitative evaluation of interferograms requires accurate interference phase measurement for any fringe position independently of field intensity variations in the reconstructed image. Unfortunately, this presents a formidable challenge to the analyst, since real interferograms are characterized by discontinuous and extraneous fringes of varying gradient and contrast, as well as an uneven background complete with speckle noise and possible shaded by anomalies in the test surface. A completely automated system would most likely include signal averaging, spatial smoothing, normalizing to local intensity values, and provisions for subtracting off the nonuniform background.

Although there is not yet a general method available that can be used to reliably interpret interferometric fringe patterns, great strides have been made to apply automated data reduction to specific applications.[30] Digital processing can consist of locating fringe centers, fitting interpolating functions, or otherwise preparing data to produce a point-by-point map of surface displacement. An alternate strategy for flaw detection is to apply pattern recognition techniques to look for anomalous holo-interferometric fringe patterns.

If fiber optic interferometry is to be commonly used as a tool in aerospace applications, some general method of simplifying the recording and analysis of complex holo-interferograms must be developed. One approach to this problem is to introduce a known carrier whose phase adds vectorally to the phase changes caused by surface deformation to produce a perturbed fringe pattern with fringe orders that monotonically increase in the direction orthogonal to the carrier. Subtraction of the carrier from the perturbed pattern aids in making peak/valley distinctions in the surface profile and unambiguously establishes the algebraic sign of the displacement vector.[31,32] This process has been implemented using digital image processing and the holographic/fiber optic system illustrated in fig. 1.[33–35]

Figure 5a shows a photograph of the top surface of a package used to house microelectronic circuit cards. This housing is made from two thin plastic shells sealed together along their edges, producing a hollow container suitable for holding an electronic card. The housing was aligned in the holo-interferometer shown in fig. 1 so that the out-of-plane displacement could be measured. The housing was fitted with an air valve and rubber hose so that surface deformations could be produced by elevating the internal pressure, and then mounted on a rotation stage to allow generation of the carrier pattern. A carrier pattern was generated on the surface of the housing by rotating the assembly around a vertical axis in the plane of the surface. Figure 5b is a photograph of the real-time fringe pattern caused by interference between the wave-

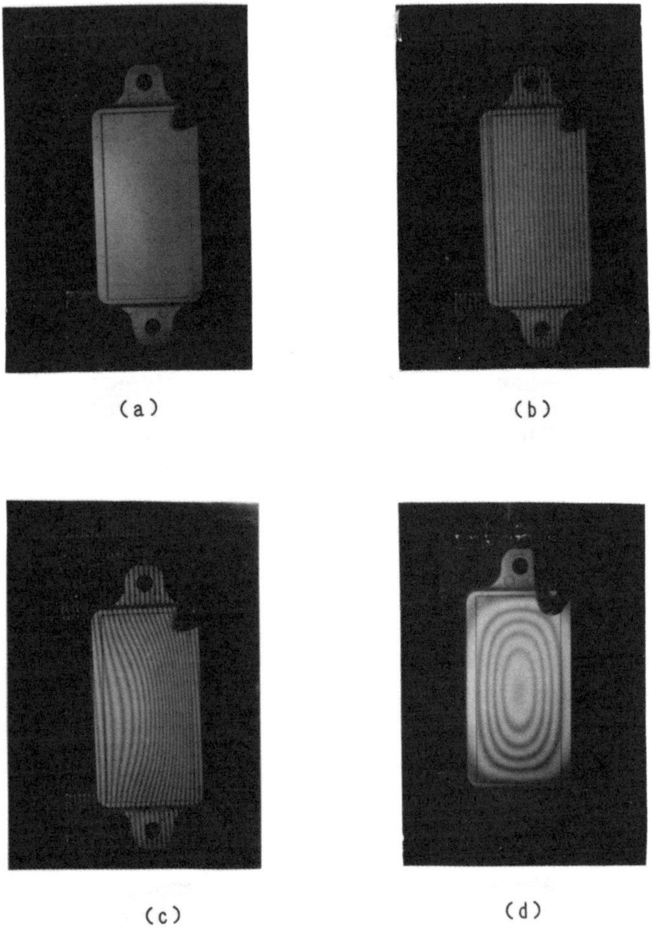

FIGURE 5- Electronic circuit housing showing (a) top surface, (b) carrier fringe pattern, (c) perturbed pattern of carrier with deformation and (d) fringe pattern showing only deformation.

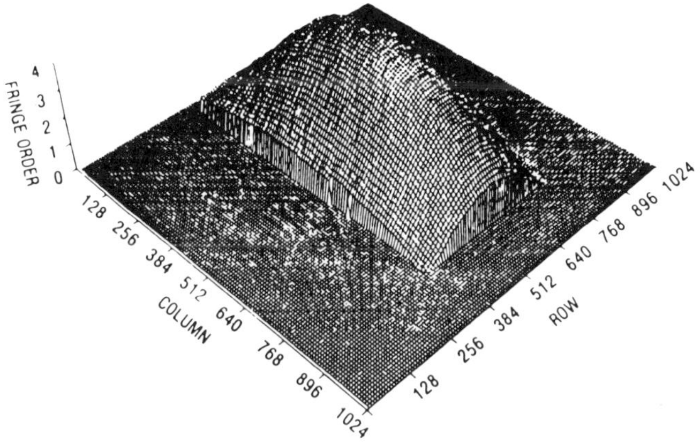

FIGURE 6- Surface contour plot of the deformed housing.

front from the holographic image of the package and the wavefront from the rotated package. The carrier pattern was modulated in real-time the internal pressure of the package was elevated. Figure 5c shows the perturbed pattern. A simple holo-interferogram of the same deformation field is shown in fig. 5d for purposes of comparison.

Since the direction of rotation of the housing is known, all the information needed to contour the deformed surface is contained in figs. 5b and 5c. The two interferometric patterns were digitized and the automated analysis was carried out on the image data files. The resultant data from the computer routines was plotted (fig. 6) to show the contour of the deformed housing.

THERMAL ACOUSTO-PHOTONIC NDE

The preceding examples illustrate that fiber optics can be used to measure surface deformations of structural components subjected to mechanical, thermal or pneumatic loadings. The authors are also working on a thermal acousto-photonic (TAP) system for nondestructive evaluation that relies on fiber-based laser excitation and sophisticated fiber optic detection. This TAP system has many potential applications to the reliability testing of structures where composition, scale, geometry, or restricted access preclude the use of conventional NDE techniques. The following section describes tests conducted to illustrate the excitation leg of the TAP system. Various means for detecting signals through optical fibers are currently being explored and the results of these tests will be reported in the near future.

The generation of acoustic waves in various materials by laser excitation has been demonstrated,[36–39] but the potential for using fiber optics for this task has been limited due to high energy densities encountered during launch at the fiber tip. Fortunately, Bjelkhagen[23,24] addressed this problem in holographic research and, from this work, the authors were able to construct a coupling device for ultrasonic NDE.[40] The device consisted of a chamber filled with index matching fluid. The optical fiber is placed within the chamber to dissipate heat energy and eliminate the air/glass interface at the fiber tip.

Figure 7 shows one of the "flawed" specimens used to demonstrate the feasibility of utilizing fiber optic excitation for nondestructive evaluation. High energy pulses from a ruby laser were launched into an optical fiber using the coupling device described above. The energy was transmitted to the specimen (an AISI 1020 steel bar containing a machined "flaw") through a 200 micron diameter multimode fiber. Rayleigh waves were monitored using a conventional piezoelectric transducer.

Figure 8a shows a reference waveform obtained by exciting a flaw free specimen using fiber excitation. This trace can be compared to the modified waveform, shown in Figure 8b, transmitted across the flaw. Higher fre-

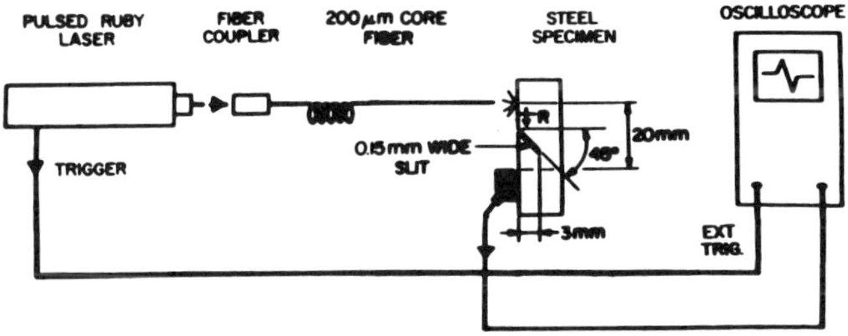

FIGURE 7- Schematic of system used to transmit a laser pulse through an optical fiber to excite a steel bar specimen which has a simulated "flaw."

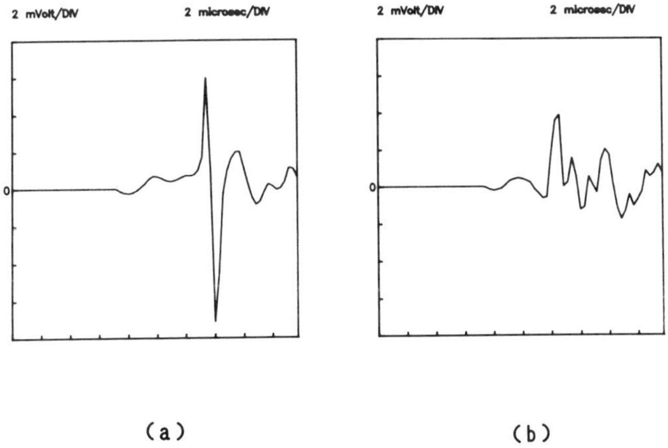

FIGURE 8- Oscilloscope traces showing the interaction between an R-wave and the steel bar specimen (a) with no "flaw" and (b) with the slanting slit "flaw."

quency components are attenuated as a function of flaw depth, and the longer pathlength around the flaw increases the time of flight. This information is sufficient to characterize the location and size of the defect.

CONCLUSION

The future of fiber optic sensors for aerospace requirements looks bright, mainly because support has been, and is being, provided for the high-risk/high-payoff ventures required to bring this area of technology to its full potential. Recent advances in equipment and image processing have overcome many of the technical difficulties associated with recording interferograms, greatly expanding the potential for incorporating interferometry into space-based diagnostic systems. Holocameras can be used to reveal deformations in real time. The resulting fringe patterns can be recorded digitally and, with the use of a carrier fringe technique, interpreted automatically. Single mode fibers can be used as stable illuminators in interferometric recording systems, and can serve as high energy light guides for thermal acousto-photonic nondestructive testing.

Acknowledgments

Portions of the research described in this paper were supported by the Alabama Research Institute under Grant No. ARI 86-303, the 1987 Fellowship Award from The American Society for Nondestructive Testing, and AT&T Bell Laboratories in Murray Hill, NJ. Dr. Gilbert's efforts in holography have been funded since 1980 by the Army Research Office in Research Triangle Park, N.C. He currently works under contract No. DAAL 03-86-K-0014.

References

1. Gabor, D., "A new microscopic principle," *Nature*, 161: 777–778 (1948).
2. Leith, E.N., Upatnieks, J., "Wavefront reconstruction with continuous tone objects," *J. Opt. Soc. Amer.*, 53: 1377–1381 (1963).
3. Rosen, A.N., "Holographic fundoscopy with fiber optic illumination," *Opt. and Laser Tech*, 7(3): 127–129 (1975).
4. Suhara, T., Nishihara H., Koyama, J., "Far radiation field emitted from an optical fiber and its applications to holography," *Trans*. of the IECE of Japan, Sec. E (Engl), 60(10): 533–540 (1977).
5. Leite, A.M.P.P., "Optical fiber illuminators for holography," *Opt. Commun.*, 28(3): 303–306 (1979).
6. Yonemura, M., Nishisaka, T., Machida, H., "Endoscopic hologram interferometry using fiber optics," *Appl. Opt.*, 28(9): 1664–1667 (1981).
7. Dudderar, T.D., Gilbert, J.A., "Fiber optic pulsed laser holography," *Appl. Phys. Lett.*, 43(8): 730–732 (1983).
8. Albe, F., Fagot, H., Smigielski, "Use of an endoscope for optical fiber holography," *Proc*. of the SPIE 2nd. Int. Tech. Sym. on Optical and Electro-Optical Applied Sci. and Eng., Cannes, France, December, (1985), pp. 199–203.
9. Gilbert, J.A., Herrick, J.W., "Holographic displacement analysis with multimode-fiber optics"', *Exp. Mech.*, 21(8): 315–320 (1981).
10. Gilbert, J.A., Schultz, M.E., Boehnlein, A.J., "Remote displacement analysis using multimode fiber-optic bundles," *Exp. Mech.*, 22(10): 398–400 (1982).
11. Gilbert, J.A., Dudderar, T.D., "Applications of fiber optics to coherent metrology for the study of material deformations and structural mechanics," *Proc*. of the Army Symposium on Solid Mechanics, Advances in Solid Mechanics for Design and Analysis, (1984), pp. 63–92.
12. Gilbert, J.A., Dudderar, T.D., Schultz, M.E., Boehnlein, A.J., "The monomode fiber-a new tool for holographic interferometry," *Exp. Mech.*, 23(2): 190–195 (1983).

13. Rowley, D., "The use of a fiber-optic reference beam in a focused image holographic interferometer," *Opt. and Laser Tech.*, 15(4): 194–198 (1983).
14. Jones, J.D.C., Corke, M., Kersey, A.D., Jackson, D.A., "Single-mode fiber-optic holography," *J. Phys. E: Sci. Instrum.*, 17: 271–273 (1984).
15. Hall, P.M., Dudderar, T.D., Argyle, J.F., "Thermal deformations observed in leadless ceramic chip carriers surface mounted to printed wiring boards," *IEEE Trans.*, CHMT-6(4): 544–552 (1983).
16. Dudderar, T.D., Hall, P.M., Gilbert, J.A., "Holo-interferometric measurement of the thermal deformation response to power dissipation in multilayer printed wiring boards," *Exp. Mech.*, 25(1): 95–104 (1985).
17. Albe, F., "Holographie ultra-rapide utilisant des fibres optiques," *Report* R 101/86, Institute Franco-Allemand de Recherches de Saint-Louis, (1986).
18. Dudderar, T.D., Gilbert, J.A., Franzel, R.A., Schamell, J.H., "Remote vibration measurement by time averaged holographic interferometry," *Exp. Tech.*, 9(1): 25–27 (1985).
19. Dudderar, T.D., Gilbert, J.A., "Real-time holographic interferometry through fiber optics," *J. Phys. Sci. Instrum.*, 18: 39–43 (1985).
20. Gilbert, J.A., Dudderar, T.D., Nose, A., "Remote deformation field measurement through different media using fiber optics," *Opt. Eng.*, 24(4): 628–631 (1985).
21. Dudderar, T.D., Gilbert, J.A., Boehnlein, A.J., "Achieving stability in remote holography using flexible multimode image bundles," *Appl. Opt.*, 22(7): 1000–1005 (1983).
22. Gilbert, J.A., Dudderar, T.D., Boehnlein, A.J., "Ultra low-frequency holographic interferometry using fiber optics," *Opt. and Lasers in Eng.*, 5(1): 29–40 (1984).
23. Bjelkhagen, H.I., "Pulsed fiber holography: a new technique for hologram interferometry," *Opt. Eng.*, 24(4): 645–649 (1985).
24. Bjelkhagen, H.I., Wesley, E.J., Liu, J.C., Marhie, M.E., Epstein, M., "Holographic interferometry through imaging fibers using CW and pulsed lasers," *Proc.* of the SPIE,

Volume 746, Industrial Laser Interferometry (1987), pp. 201–208.
25. Post, D., "Moire interferometry for deformation and strain studies," *Opt. Eng.*, 24(4): 663–667 (1985).
26. Basehore, M.L., Post, D., "Displacement fields (u,w) obtained simultaneously by moire interferometry," *Appl. Opt.*, 21(14): 2558–2562 (1982).
27. Post, D., Czarnek, R., Joh, D., "Shear strains in a graphite/PEEK beam by moire interferometry with carrier fringes," *Exp. Mech.*, 27(3): 246–249 (1987).
28. Post, D., "Moire interferometry at VPI & SI," *Exp. Mech.*, 23(6): 203–210 (1983).
29. Johnson, H.S., Gilbert, J.A., Matthys, D.R., Dudderar, T.D., "Real-time moire interferometry," accepted for publication in *Exp. Mech.* (1987).
30. Trollinger, J.D., "Automated data reduction in holographic interferometry," *Opt. Eng.*, 24(5): 840–842 (1985).
31. Gilbert, J.A., Dudderar, T.D., Matthys, D.R., Chern, J.H., "Digitization of holographic interferograms for deflection measurement," *Proc.* of the 1985 Society for Experimental Mechanics Spring Conference on Experimental Mechanics, Las Vegas, Nevada, June 9–13, 1985, pp. 294–301.
32. Plotkowski, P.D., Hung, Y.Y., Hovanesian, J.D., Gerhart, G., "Improved fringe carrier technique for unambiguous determination of holographically recorded displacements," *Opt. Eng.*, 24(5): 754–756 (1985).
33. Matthys, D.R., Gilbert, J.A., Dudderar, T.D., "Computer analysis of holo-interferograms," *Proc.* of the Spring Conference on Experimental Mechanics, Society for Experimental Mechanics, New Orleans, LA, June 1986, pp. 833–838.
34. Matthys, D.R., Dudderar, T.D., Gilbert, J.A., "Automated analysis of holo-interferograms for the determination of surface displacement," accepted for publication in *Exp. Mech.*.
35. Matthys, D.R., Gilbert, J.A., Dudderar, T.D., Koenig, K.W., "A windowing technique for automated analysis of holo-interferograms," accepted for publication in *Opt. and Lasers in Eng.* (1987).

36. Scruby, C. B., Dewhurst, R. J., Hutchins, D. A. and Palmer, S. B., "Laser generation of ultrasound in metals," *Research Tech. in Nondestructive Testing*, 5: 281–327 (1982).
37. Birnbaum, G. and White, G. S., "Laser techniques in NDE," *Research Tech. in Nondestructive Testing*, 259–365 (1984).
38. Hutchins, D. A. and Tam, A. C., "Pulsed photoacoustic materials characterization," *IEEE Trans*, UFFC-33(5): 429–449 (1986).
39. Hutchins, D. A., "Ultrasonic generation by pulsed lasers," *Physical Acoustics*, Vol. 18, W. P. Mason & R. N Thurston, Eds., Academic Press, NY (1986).
40. Burger, C.P., Dudderar, T.D., Gilbert, J.A., Peters, B.R., Smith, J.A., "Laser excitation through fiber optics for NDE," *J. of Nondestructive Evaluation*, 6(1): 57–64 (1987).

Use of Digital Image Correlation for Determination of Displacements and Strains

H. A. Bruck, S. R. McNeill, S. S. Russell, and M. A. Sutton

College of Engineering,
University of South Carolina,
Columbia, South Carolina 29208

INTRODUCTION

There is an increasing interest in mechanics directed towards the development of nondestructive techniques for aerospace requirements. Many of these requirements necessitate knowledge of both the material and its behavior. For example, it is often necessary to know: (a) the failure and yield stresses, (b) fracture parameters, (c) various material constants for composite materials, and (d) the current damage state for the material. Techniques used to determine these parameters include ultrasonic inspection, radiography, moiré interferometry, and automated hybrid techniques. Recently, a techniquc employing an optical digitization system with digital image correlation has been developed at the University of South Carolina.[1,2] This technique accurately measures planar translational motion using bilinear interpolation and a optimization technique which uses a coarse-fine search methodology.

Recently, this technique was modified to include a Newton-Raphson search method,[3] bicubic spline interpolation[4] of the intensity data, and averaging of images to efficiently and accurately measure both rotational and displacement components.[5] Although the modified technique can predict strain and displacement components, only limited experimental work has been carried out to test this technique's accuracy and efficiency in measuring strain and displacement components.

The following work quantifies the inherent accuracy and efficiency of the aforementioned digital correlation techniques in predicting one-dimensional strains and displacements. Comparisons are made between (a) the actual strain, (b) the predicted strains using the gradients of displacements directly determined by the Newton-Raphson method, and (c) the strains obtained from curve fits to the displacement data from the coarse-fine search method.

DISCUSSION OF GENERAL METHOD

The foundation of these methods is digital image processing. Digital image data of a planar object is acquired with a video camera that has an array of sensors located on the image plane of the camera. The sensors measure discrete samples of light intensity over a short period of time ($\frac{1}{60}$th of a second) from many areas of the object plane. Figure 1 portrays a schematic of a typical system used to record digital intensity data. The discrete samples (called pixels) are then assigned an eight bit number representing the average light intensity (gray level) over time. Black is represented by 0 and white by 255. Therefore, the digitized image that is obtained is not a continuous image, but a matrix of numbers. Figure 2 shows a typical set of intensity data for an 8×8 subset of pixels. If each point is said to represent the intensity of light for a finite area, then one could reconstruct a continuous representation of the object plane's intensity distribution using interpolation methods. Figure 3 presents the re-

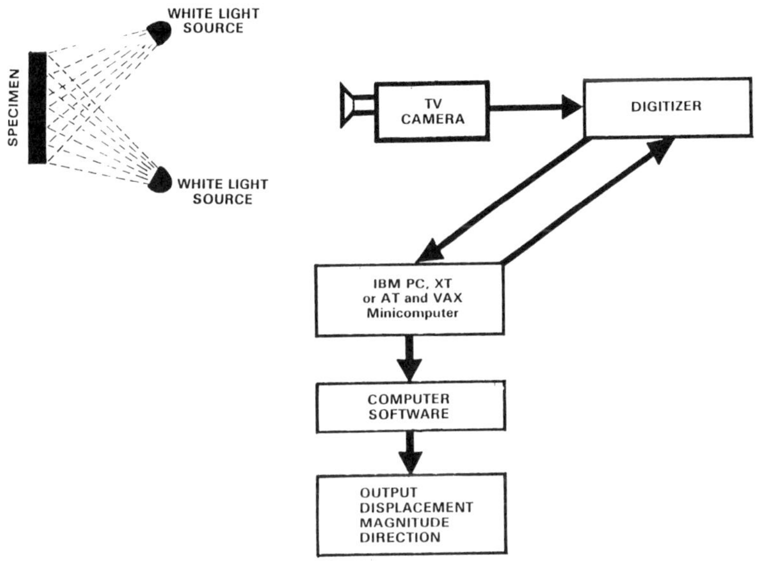

FIGURE 1- Schematic diagram of a typical system.

constructed form for the same 8 × 8 area using bicubic spline interpolation.[4]

One method commonly used to determine the deformations of a subset is digital image correlation. Digital image correlation is applied by mathematically comparing two digitized images of an object. One image is of the object before deformation, and it is referred to as the undeformed image. The other image is obtained after the object has undergone translational, rotational, and/or strained motion relative to the camera, and is referred to as the deformed image. Using these two images, in-plane deformations between images can be mathematically deduced by correlating two subsets. One subset is composed of the intensity values of a chosen number of points

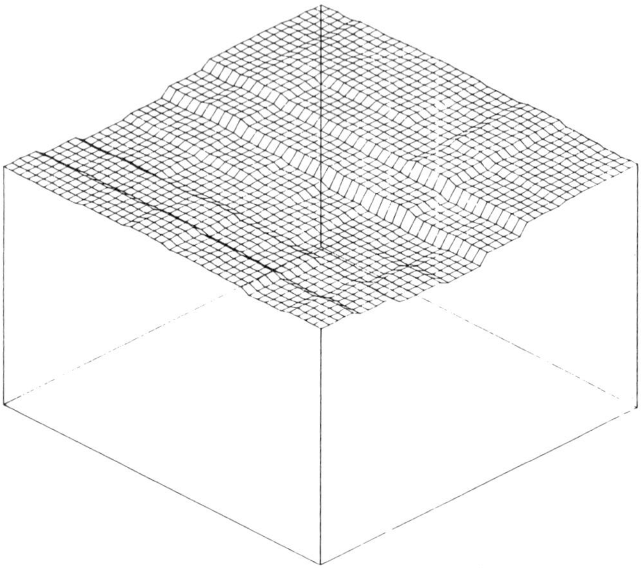

FIGURE 2- Uninterpolated 8 × 8 subset of image data.

on the undeformed image. The second subset is composed of the intensity values of each undeformed point's corresponding deformed location on the deformed image. Typically, these subsets are on the order of 20 pixels by 20 pixels.

The correlation function chosen for this work is the cross-correlation function, S, defined as:

$$S\left(u,v,\frac{\partial u}{\partial x},\frac{\partial v}{\partial y},\frac{\partial u}{\partial y},\frac{\partial v}{\partial x}\right) = 1 - \frac{\sum\sum (F(x_i,y_j)*G(x_i^*,y_j^*))}{\sqrt{\sum\sum F(x_i,\, y_j)^2 * \sum\sum G(x_i^*,y_j^*)^2}} \quad (1)$$

where the double summation within the correlation func-

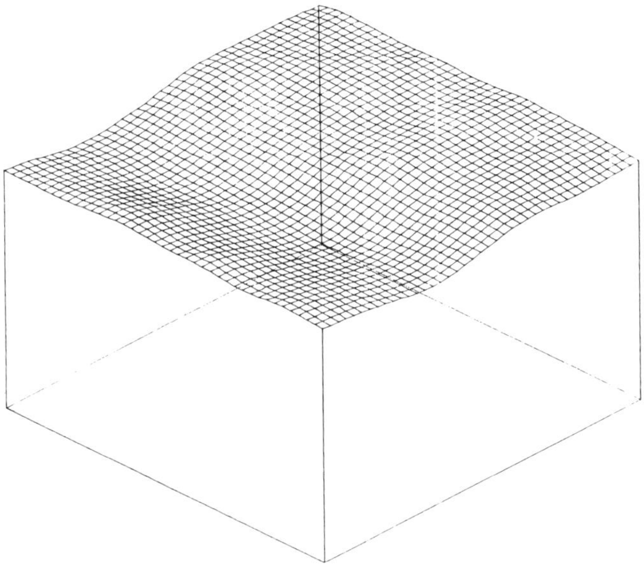

FIGURE 3- 8 × 8 subset of image data with bicubic spline interpolation.

tion is over the entire subset. $F(x_i,y_j)$ represents the intensity value at coordinates (x_i,y_j) on the undeformed image, and $G(x_i^*,y_j^*)$ represents the intensity value at coordinates (x_i^*,y_j^*) on the deformed image. Through deformation analysis and Taylor series expansion, the coordinates (x_i^*,y_j^*) can be approximated in terms of the deformation parameters as follows:

$$x_i^* \simeq x_i + u + \frac{\partial u}{\partial x} \cdot \Delta x + \frac{\partial u}{\partial y} \cdot \Delta y$$

$$y_i^* \simeq y_i + v + \frac{\partial v}{\partial y} \cdot \Delta y + \frac{\partial v}{\partial x} \cdot \Delta x \qquad (2)$$

where u and v are the displacements of the subset's center. The displacement gradients

$$\left(\frac{\partial u}{\partial x}, \frac{\partial v}{\partial y}, \frac{\partial u}{\partial y}, \frac{\partial v}{\partial x}\right)$$

are assumed to be constant over the subset. The terms Δx and Δy are the relative distance from the center of the subset to the point of interest in the subset.

If interpolation is not used, the values x_i, y_j, x_i^*, and y_j^* must be integer values. Consequently, the six deformation parameters calculated would not be independent, but would have to satisfy Eq. (2). If deformations of the object occurred which did not fit the above restriction, the values of the deformation parameters predicted by minimization of Eq. (1) would be in error. In some fields of digital image analysis, the errors in the deformation parameters would be perfectly acceptable. In the field of experimental stress analysis, the predicted values are virtually useless. It then becomes necessary to implement a method to acquire intensity values at sub-pixel locations so that independent deformation parameters can be calculated. In this work bicubic spline functions[4] were used to approximate sub-pixel intensity values for the Newton-Raphson method, and bilinear interpolations were used in the coarse-fine search method.

In summary, the process of determining displacements and displacement gradients for many positions on an object surface using digitized images of the object involves the following steps:

(a) Place a randomly varying intensity pattern on the object's surface of the "appropriate" spatial frequency.
(b) Digitize the undeformed and deformed positions of the object with a CCD camera interfaced to a microcomputer. Store the images on the computer.

(c) Process the images using both intensity fitting algorithms and digital correlation to determine the displacements and displacement gradients for the subsets of interest.
(d) Process the deformation data to estimate the strain field for the object.

EXPERIMENTAL ANALYSIS

A polycarbonate specimen, with dimensions shown in Figure 4, was used in the experiments. The front surface of the specimen was sprayed with fine mists of black and white paint to obtain a randomly varying intensity pattern. A typical intensity pattern is shown in Figure 5. The polycarbonate specimen was placed in an MTS tension-compression materials test machine and fixed into position with a pair of mechanical wedge grips. Both the MTS machine and image acquisition equipment used in this experiment can be seen in Figure 6.

A baseline test was performed to determine the inherent error in the correlation procedure. This test used two images of the specimen with no deformation of the specimen. Images used in the correlation procedure are the average of twenty images. This was done to minimize the effects of electronic noise associated with the digitization process. The specimen was subjected to two tensile loads of different magnitude. Images of the object and extensometer measurements were taken for each loading.

The image acquisition system used in this work includes a Sony CCD XC-38 video camera with a 50mm lens and a 25mm lens with extension rings. A typical magnification factor for the images was 200 pixels/cm. A video digitizing board, Matrox Incorporated PIP-1024, was used for digitizing and displaying images. Data was stored on floppy disks and later transferred to a VAX 11/

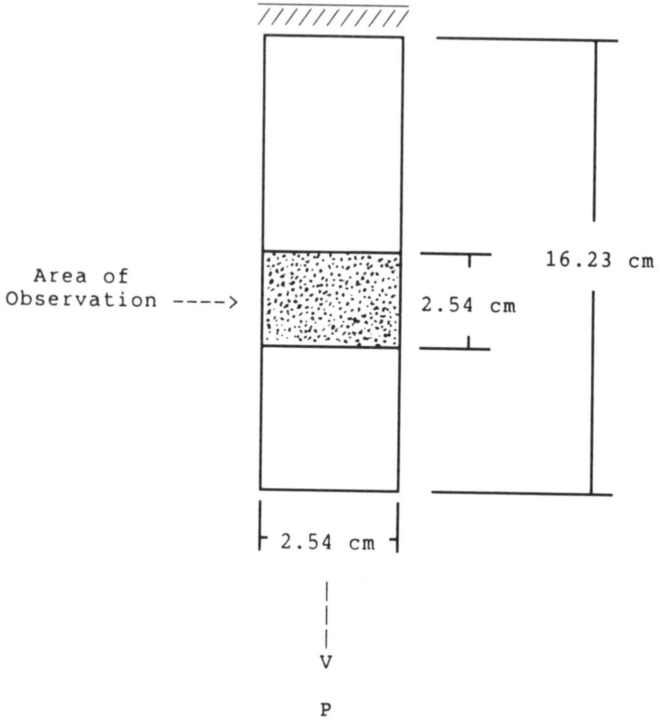

FIGURE 4- Polycarbonate specimen used in testing experiment.

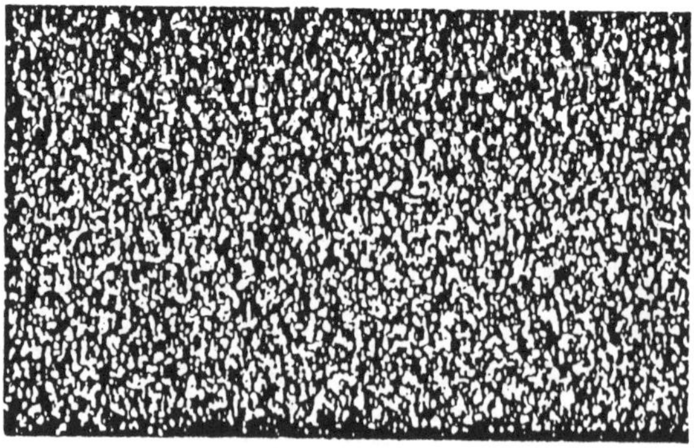

FIGURE 5- Random pattern for baseline test.

FIGURE 6- MTS machine and image acquisition equipment used in experiment.

780 super minicomputer through the serial port of an IBM PC/AT.[6]

Analysis of the images to obtain estimates of the strains was accomplished using two numerical optimization programs. The first method of optimization uses a coarse-fine search method to determine *only* the displacements of the subset's center assuming the gradients of displacement for an individual subset are zero. Gradients were determined from the displacement data using both finite difference and least squares curve fitting methods. The second method uses a Newton-Raphson search procedure to determine the displacements and displacement gradients for each subset.

Subset sizes were 20×20 or 50×50 for the correlation work. All averages and standard deviations for the predicted experimental results are computed from 100 subsets within a 100×100 pixel area centered in the image plane.

DISCUSSION OF RESULTS

Table 1 presents a comparison of the execution times for two programs; COLUMN, which uses a coarse-fine search method, and BICUBIC, which uses the Newton-Raphson search method. Table 2 presents a comparison of (a) the actual strain, (b) the strain predicted from the gradients obtained from the BICUBIC program, and (c) the strain calculated from the curve fits to the displacement data obtained from the COLUMN program. It is evident from inspection of Table 2 that a least squares fit to the displacement data obtained from COLUMN yields results that have an absolute error in strains of approximately 0.004% with a standard deviation of 0.049% for strains of 0.5%. Similarly, for strains of 1.0% COLUMN yields results that have an absolute error of 0.037% with

TABLE 1-

Average Time per Subset (in seconds*)			
Newton-Raphson Method		Coarse-fine Search Method	
50 × 50 Subsets	20 × 20 Subsets	50 × 50 Subsets	20 × 20 Subsets
20.25	4.04	176.86	27.96

* CPU seconds on a VAX 11/780 super minicomputer

TABLE 2-

Strain (du/dx) for Uniaxial Loading Averaged Over 40 Subsets				
	Newton-Raphson Method		Coarse-fine Search Method (50 × 50 subsets)	
Actual	50 × 50 Subsets	20 × 20 Subsets	Finite Difference	Least Squares
0.0% ± 0.005%	0.028% ± 0.014%	0.088% ± 0.104%	0.014% ± 0.069%	−.005% ± 0.008%
0.5% ± 0.005%	0.534% ± 0.023%	0.653% ± 0.213%	0.510% ± 0.027%	0.504% ± 0.049%
1.0% ± 0.005%	1.066% ± 0.035%	1.181% ± 0.228%	1.031% ± 0.024%	1.037% ± 0.099%

a standard deviation of 0.099%. The finite difference methods gave a slightly lower accuracy with an absolute error of approximately 0.010% for strains of 0.5% and 0.031% for strains of 1.0%, but the average standard deviation was 0.027% which is slightly better than the least squares method. The BICUBIC program determined the displacement gradients directly with an absolute error of approximately 0.034% for strains of 0.5% and 0.066% for strains of 1.0% with an average standard deviation of 0.023%.

The computer time required to correlate 50 × 50 subsets using COLUMN and BICUBIC were noticeably different, with BICUBIC exhibiting a speed nearly an order of magnitude greater than COLUMN. This does not include the time needed to execute the curve fitting routines to the displacement data obtained with COLUMN. In summary, BICUBIC was more efficient, but less accurate than COLUMN in determining strain.

Relative to sources of error in correlating data from this experiment, it is important to note that there were problems not mentioned in Ref. [5] that were unique to the testing machinery that was used. Most notably was a slight out-of-plane rotation induced in the specimen by the grips. It was deduced that this problem, although trivial for the given deformations, did contribute to some of the experimental error. However, it was not possible to quantify this effect.

CONCLUSION

The Newton-Raphson method has been shown to be accurate and efficient in directly predicting the actual values for the strain and associated displacements of individual subsets between undeformed and deformed images. The coarse-fine search method was found to be slightly more accurate in predicting strains than the Newton-Raphson method, but the technique was far less efficient and required curve fits to the displacement data in order to predict strains.

References

1. Peters, W H, Ranson, W F and Sutton, M A *et al.* 'Applications of digital image correlation techniques to experimental mechanics' *Experimental Mechanics* Vol 22 No 6 (1983) pp 232–244.
2. Sutton, M A, Wolters, W J *et al.* 'Computer vision determination of displacement using an improved digital correlation method' *Image and Vision Computing* Vol 1 No 3 (1983) pp 133–139.
3. Sutton, M A, Cheng, Mingqi, Peters, W H, *et al.* 'Application of an optimized digital correlation method to planar deformation analysis' *Image and Vision Computing* Vol 4 No 3 (1986) pp 143–150.
4. De Boor, Carl. 'Bicubic spline interpolation' *Journal of Mathematics and Physics* Vol 41 (1962) pp 212–218.
5. McNeill, S R., Bruck, H A, Sutton, M A. 'An Improved Method for Correlating Digital Images Using the Newton-Raphson Method and Bicubic Splines' To be published in *Experimental Mechanics*.
6. McNeill, S R. Private Correspondence.

Optical Generation and Detection of Ultrasound with Acousto-Ultrasonic Applications

James W. Wagner and Robert E. Green, Jr.

Center for Nondestructive Evaluation,
The Johns Hopkins University,
Baltimore, Maryland 21218

ABSTRACT

The potential for generating broad bandwidth acoustic signals and detecting them with extreme fidelity makes optical methods attractive for acousto-ultrasonic testing applications. Laser generated acoustic signals may be detected using any one of several interferometer configurations which have been shown to offer theoretical displacement sensitivities on the order of 10^{-5} Angstroms per square root Hertz of system bandwidth. Research is continuing in efforts to improve detection sensitivity while affording greater immunity to specimen surface roughness. Likewise, laser generation schemes are under development which will improve the efficiency with which acoustic waves can be coupled into a material and detected after propagation.

INTRODUCTION

Optical techniques for performing ultrasonic testing functions offer several advantages over traditional piezoelectric techniques. Chief among them is the potential for performing ultrasonic testing without contacting the surface of the specimen being investigated. The fact that it is not necessary to use any form of mechanical couplant between the surface and the generating or detecting transducer offers both practical and technical advantages as well. For example, acoustic attenuation measurements may be made with much greater accuracy since they are no longer affected by variations in coupling from point to point on the object surface. Without couplant, the ease with which a surface may be scanned is greatly enhanced, as well. Since optical beams may be focused to very small spot sizes, complex surface geometries may be scanned with high spatial resolution relative to conventional contacting transducer techniques. Finally, optical techniques offer the possibility of generating extremely broad band signals and recording the effects of their propagation through a specimen with high fidelity.

OPTICALLY GENERATED ACOUSTIC WAVES

It is the broad band nature of optical techniques applied to ultrasonic testing which may be used to particular advantage in acousto-ultrasonics. Consider first the characteristics of optically generated ultrasound. The output of a solid state pulsed laser may give rise to a monophasic acoustic signal tens of nanoseconds in pulse width. The spectral content of such an ultrasonic pulse extends over a frequency range from dc and may approach 100 MHz. On the other hand the output from a contacting piezoelectric transducer is limited by the natural resonance of the piezoelectric crystal. Even a highly damped transducer which generates only a single cycle of RF energy has a corresponding power spectrum with very little energy at low frequencies and a upper bandwidth of only a few megahertz. In acousto-ultrasonic applications where the effect of the material and it's defect structure on acoustic wave propagation may be a strong function of spectral content, the broad band laser generated signal has a distinct advantage relative to contacting transducers.

The mechanism by which a laser generates an acoustic wave in a material may be modeled in several ways. Perhaps the most popular model[1] is one in which the laser is assumed to cause a rapid thermal expansion in a small volume of the material near its surface. Propagation of the acoustic wave as a result of a sudden increase in material volume is then computed based on the mechanical properties of the material. More complete models of the acoustic wave generation process include coupling between the propagating thermal energy and the acoustic energy.[2] However the simpler model, which ignores thermal coupling, has been shown to give excellent results when compared with experimental data.[3]

Three cases for laser generation of acoustic waves are usually considered—free surface, constrained surface, and ablative surface cases. These three cases are illustrated in fig. 1. When a laser beam strikes the free surface of a material as shown, the heat absorbed by the material near the surface results in a localized expansion. Since the material surface is not constrained, a significant portion of the expanded volume is free to move away from the bulk of the material and into the air adjacent to the material surface. As a result, most of the acoustic energy generated does not propagate on-axis, normal to the material surface but instead propagates in a cone with an apex half angle of about 60 degrees as shown in the figure. Higher on-axis acoustic energy may be generated by constraining the surface from bulging into the air when it is heated. This may be done mechanically by placing a glass cover slip over the region to be illuminated by the laser, for example. Alternatively one may coat the surface with a transparent liquid. An even stronger on-axis source may be obtained by placing a material on the specimen surface which may be ablated by the laser beam or alternatively by using sufficient beam energy to ablate the surface of the material itself. It is in fact this later ablative case

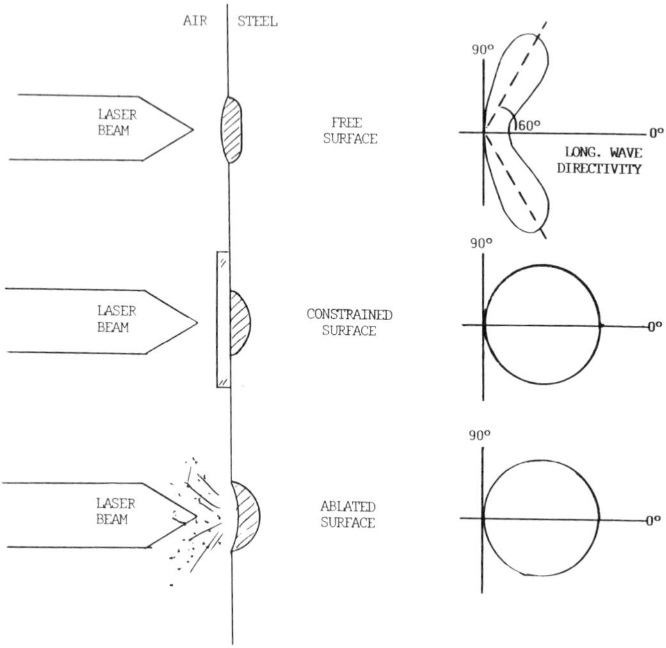

FIGURE 1- Laser generation of acoustic waves.

where the greatest acoustic energy is obtained. It should be noted, however, that only in the first case, that is the free surface thermo-elastic generation case, is laser generation of acoustic waves truly noncontacting and nondestructive.

OPTICAL DETECTION OF ULTRASOUND

While several approaches to optical detection of acoustic waves in materials have been proposed and described in the literature,[4–7] interferometric techniques are of greatest utility in broad band detection of acoustic waves. The most common forms of interferometers for these applications fall into three categories—path stabilized, heterodyne, and wavelength selective. The path stabilized interferometer is illustrated in fig. 2. In this case the interferometer is basically a Michelson type system where the position of the mirror in the reference leg may be changed by the displacement of a piezoelectric translator. In order that the interferometer may be maintained in its most sensitive configuration, a signal from the detector is processed to control the reference beam pathlength.

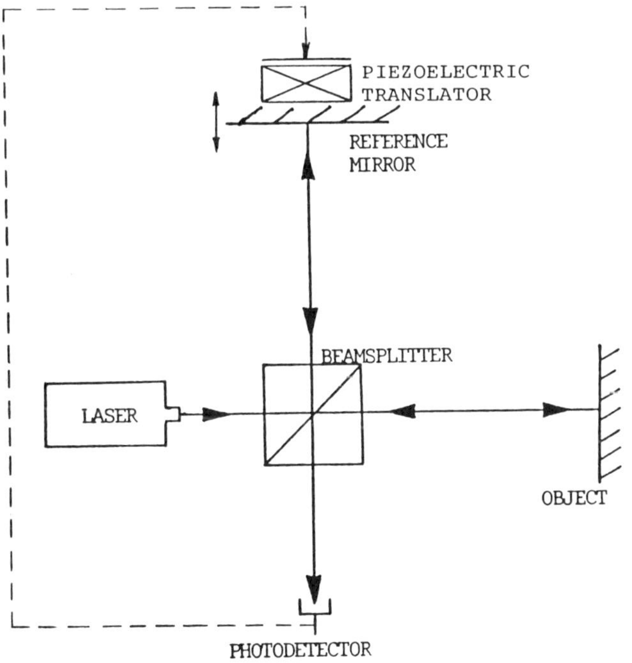

FIGURE 2- Path-stabilized interferometer.

Using such stabilization, the interferometer can be made immune to ambient vibration or temperature changes which might otherwise alter optical pathlengths and degrade the performance of the system. Of the three interferometric systems, path stabilized interferometers are the most sensitive by a small margin but also have the greatest requirement for high specimen surface reflectivity.[8]

Heterodyne interferometry is an alternative to path stabilization for providing immunity to environmental noise sources. The basic difference between heterodyne and path stabilized (homodyne) interferometry is that the two optical wavefronts which interfere at the detector of the interferometer are shifted in frequency relative to each other in the heterodyne case. A configuration employing heterodyne techniques is illustrated in fig. 3. In this case the interferometer is configured using optical fibers rather than the bulk components. The frequency shift necessary to perform heterodyning is provided by an acousto-optic cell through which the laser beam is made to pass. When properly aligned, light from the cell exists at two slightly different angles with the on-axis or zero-order beam passing straight through the cell. The frequency of this beam is unshifted from that of the original laser source. The second beam or first order diffracted beam, however, exits the cell at a slight angle relative to the zero-order beam and is shifted in frequency by 50 to 100 MHz typically. In the fiber configuration shown in the figure, unshifted light propagates clockwise around the primary loop or fiber while the shifted light propagates in the counter-clockwise direction around the same loop. The shifted light is coupled across the fiber optic coupler and exits the free end of the dfiber where it is focused on the specimen surface. Once reflected by the specimen surface, the light is refocused into the fiber and propagates back through the second fiber to a photodetector. The un-

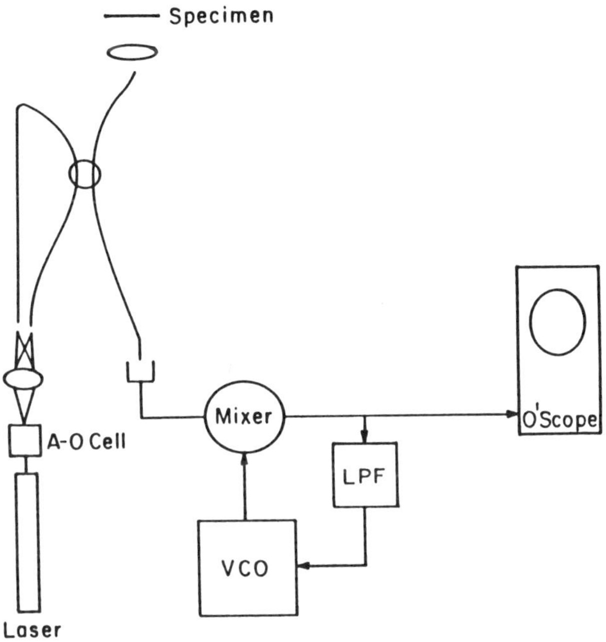

FIGURE 3- Fiber-optic heterodyne interferometer.

shifted light from the laser, which propagates in a clockwise direction around the primary loop, is also coupled into the second fiber but in this case goes directly to the optical detector. Therefore, both the shifted and unshifted light beams arrive simultaneously at the photodetector where they interfere to generate an intensity signal which is an offset sinusoid whose intensity varies at the difference shift frequency. When used in this configuration, a dynamic displacement of the specimen surface results in a phase shift in the detected electronic signal. One may detect this phase shift using a phase locked loop configuration as shown in the figure. In this case an incoming signal is mixed with a local oscillator signal which is controlled in order that its output remains in quadrature with the optical signal. A low pass filter with a cut-off frequency of several kilohertz is inserted in the loop in order that the loop will remain stabilized only for low frequency variations such as those associated with environmental disturbances. High frequency fluctuations, which may be assumed to be acoustic signals, are not compensated by the loop and therefore pass directly to recording and analysis electronics. The heterodyne technique has a theoretical displacement sensitivity only slightly less than the path stabilized configuration.[8] Furthermore, the technique is more tolerant of variations in specimen surface reflectivity.

Perhaps the least stringent requirement for specimen surface flatness and reflectivity is offered by the frequency selective class of interferometric detectors. The Fabry-Perot interferometer, for example, has been used very successfully in this application[9] and is illustrated in block diagram form in fig. 4. Actively tuned to the wavelength of the laser beam illuminating the specimen surface, a large aperture interferometer is used to collect as much of the light scattered by the specimen as possible. Note from the figure that the specimen surface may be rough and optically diffuse. When the specimen surface moves in response to acoustic excitation, the frequency of the light scattered by the surface is doppler shifted so that its wavelength is changed slightly. Since the Fabry-Perot cavity is tuned relative to the laser source wavelength, any variation in the wavelength of the light passing through the cavity causes a change in the signal intensity at the output detector.

In spite of the differences in their principles of operation, all three classes of interferometers have quite similar ultimate sensitivity specifications.[7] Assuming that the ultimate limits of displacement sensitivity result from the

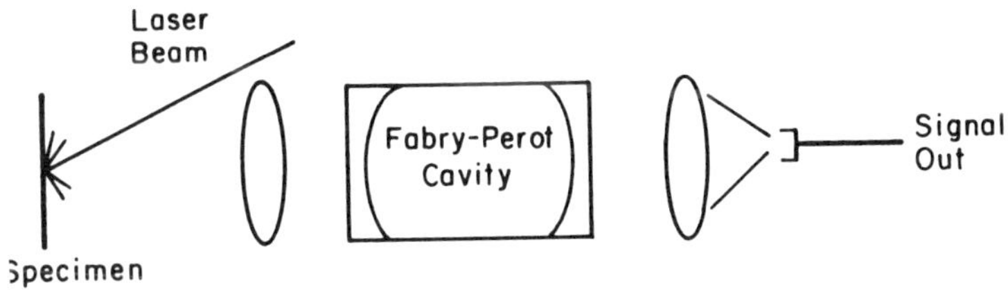

FIGURE 4- Stabilized Fabry-Perot interferometer.

statistical uncertainties associated with light wave propagation, then the minimum detectable surface variation is governed in all three cases by the proportionality below:

$$\text{min detectable variation} \propto \left[\frac{\text{system bandwidth}}{\text{power on detector}}\right]^{1/2}$$

From this expression it is clear that improvements in sensitivity may be obtained either by increasing the amount of light which falls on the optical detector or by reducing the electronic bandwidth of the system. It is for this reason that at least one commercial manufacturer of laser generation and detection systems uses, in the detection system, a pulsed laser that provides a long pulse of higher energy than is produced by the laser used to generate acoustic waves. For typical values of laser beam power and detector efficiencies used in laboratory systems, the ultimate sensitivity for all three classes of interferometers is in the neighborhood of 10^{-5} angstroms per square root hertz of system bandwidth. Thus for a system bandwidth of 10 MHz, for example, the ultimate displacement sensitivity would be about .03 angstroms. At current levels of sensitivity, interferometric systems are in general less sensitive than most contact transducer systems. Thus for very weak acoustic signals, it is often necessary to operate interferometric systems in a narrow bandwidth mode and to scan the center frequency of the system over a broad range of frequencies, repeating the tests at each new frequency.

CONCLUSIONS

Owing to their broad bandwidth and independence from coupling variations, optical techniques for generating and detecting acoustic signals offer the means by which to perform acousto-ultrasonic testing with extreme fidelity. Interferometric systems continue to be developed to provide improved sensitivity. New designs incorporating nonlinear, phase conjugating components have been proposed which should insure greater immunity to surface roughness and reflectivity.[10] Similarly, efforts to improve the efficiency with which acoustic waves may be coupled into test specimens from laser sources are continuing as well. Through these combined efforts it is hoped that practical optical systems for generating and detecting acoustic signals will be available in the very near future for a wide range of applications.

References

1. Rose ORF, Point-source representation for laser-generated ultrasound, J. Acoust. Soc. Am. *75*(3), 723–732 (1984).
2. Schleichert U et al., Theoretical and experimental investigation of broadband thermoelastically generated ultrasonic pulses, Proc. Conference on Photoacoustics, Heidelberg, Germany, (1987).
3. Scruby CB, RJ Dewhurst, DA Hutchins, and SB Palmer, Quantitative studies of thermally generated elastic waves in laser-irradiated metals, J. Appl. Phys. *51*(12), 6210–6216, (1980).
4. Whitman RL and A Korpel, Probing of acoustic surface perturbations by coherent light, Applied Optics *8*(8), 1567–1576, (1969).

5. De La Rue RM et al., Acoustic-surface-wave amplitude and phase measurements using laser probes, Proc. IEE *119*(2), 117–126, (1972).
6. Palmer CH and RE Green, Jr., Optical probing of acoustic emission waves, in *Nondestructive Evaluation of Materials*, edited by JJ Burke and V Weiss (Plenum Press, New York, 1979), 347–378.
7. Monchalin J-P, Optical detection of ultrasound, IEEE Trans UFFC, Vol. UFFC-*33*(5), 485–499, (1986).
8. Wagner JW and JB Spicer, Theoretical noise-limited sensitivity of classical interferometry, J. Opt. Soc. Am. B *4*(8), 1316–1326, (1987).
9. Monchalin J-P, Optical detection of ultrasound at a distance using a confocal Fabry-Perot interferometer, Appl. Phys. Lett. *47*(1), 14–16, (1985).
10. Paul M, B Betz, and W. Arnold, Interferometric detection of ultrasound at rough surfaces using optical phase conjugation, Appl. Phys. Lett. *50*(22), 1569–1571, (1987).

Overview of Non-Destructive Evaluation (NDE) Technologies for Space

Joe Sciabica

Air Force Astronautics Laboratory, Edwards AFB, California

ABSTRACT

A critical need will exist within the space materials community to non-destructively inspect and evaluate structures and components in space so that accurate predictions of system performance can be made. This paper will provide an overview of the current NDE technologies for propulsion and space applications. The contents of the paper are based on an informal survey of industry and government agencies carried out by personnel from the Air Force Astronautics Laboratory (AFAL) during the summer of 1987.

OVERVIEW

The AFAL is located in the Mojave Desert at Edwards Air Force Base, CA. The Laboratory is the principal Air Force Systems Command organization charged with planning and executing the United States Air Force research, exploratory development, and advanced development programs for interdisciplinary space technology and rocket propulsion. Space technology includes spacecraft technology, launch vehicle technology, launch and in-space operations, and logistics technology. During the summer of 1987, an informal survey of industry and government agencies was performed by the AFAL. This survey included the area of NDE for space structures prior to launch and after space basing. NDE will play a major role in space structures because of interests in 20 year mission life of space structures. A goou NDE inspection plan must first be implemented to insure that the components are being fabricated to required specifications. In fig. 1 we see such a plan with six major parts, in which the operations are a mixture of earth and space-based inspections. Therefore, NDE technologies need to be developed for operation in both environments to yield directly comparable quantitative NDE data.

For example, to determine if a space structure has suffered damage or degradation since it was launched into space, a quantitative data base would be required to determine the extent of any damage or degradation that has

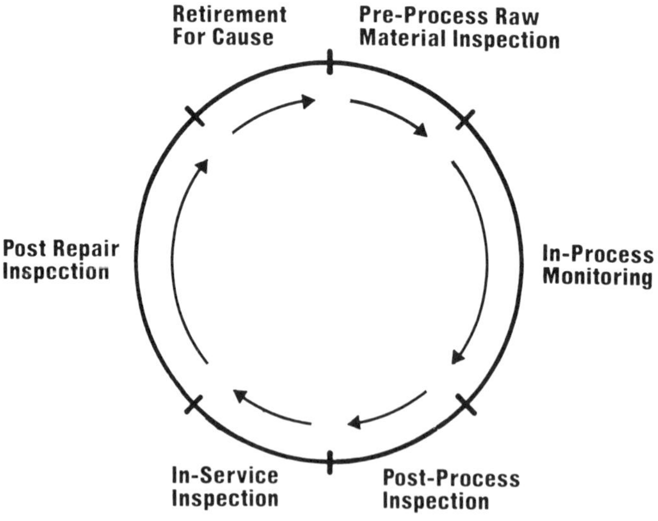

FIGURE 1- NDE for component life cycles.

occurred. The NDE inspection would enable the determination of whether on-orbit repairs are required or if the space structure should be retired. If on-orbit repairs are required, an NDE technique should be available that would be capable of assuring that the repair job was performed adequately and that the space structure is again operable. If it is decided to retire the space structure, it is necessary to know when the damaged parts would fail so that a new structure could be launched, people and equipment removed from the old structure, and the old structure disposed of properly.

The area of materials also poses a challenge for NDE inspection. Conventional space materials currently are made up of monolithic materials such as aluminium and titanium alloys. The push in new space structure materials is leaning toward fiber reinforced components consisting of ceramic matrix, metal matrix, and polymer matrix composites with coatings and hybrid resins. The designers are being driven to these fiber reinforced composites mainly for their inherent stiffness and low weight. These fiber reinforced composites present the inspector with a new challenge because by nature they are full of cracks, voids, and inclusions. These same items found in metallic components would warrant component rejection. However, in the composite component these items are important to the component; without them the component would fail. There are two types of flaws that the inspector must be able to find: isolated and distributed. The isolated flaws include voids, inclusions and cracks; the distributed flaws include bonding, delaminations, porosity, and fiber management. Detection of these flaw types becomes a function of the inspection technique and the amount of surface area inspected. For example, on a solid rocket motor the prevalent method of case to insulator to propellant bondline inspection is the use of tangential x-ray. Tangential x-ray shots are taken every 30 degrees throughout the motor. Figure 2 shows the surface area missed by performing this type of inspection. As the motor is rotated and x-rayed, only 30–60 percent of the volume is inspected. Hence it is very easy to miss an isolated flaw with this inspection method. Due to the time, cost, and amount of film required, this type of inspection is prohibitive on a large scale for production type motors. Also, the fact remains that tangential x-ray is a qualitative NDE technique subject to the interpretation of the film reader. The potential exists for the film reader to miss an isolated flaw that is picked up on film. Therefore, as we examine NDE techniques for space structures, we will consider techniques with greater than 60 percent quantitative volume inspection capability.

The measurable variables must be determined as NDE equipment is developed for space structures. These can be divided into three main areas; structural, physical, and environmental. Structural variables include strain, bending, crack growth, thickness change, and part movement (accelerations and vibrations). Physical variables include

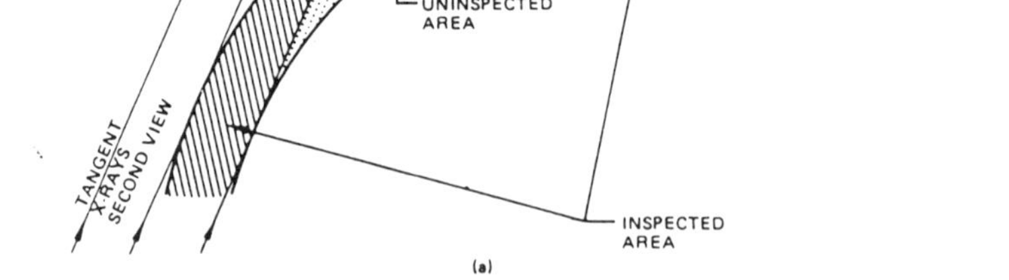

FIGURE 2- Uninspected areas due to x-ray technique.

temperature, pressure, density, change in mass, and visual appearance. Environmental variables include impact damage, atomic oxygen, radiation, magnetic fields, and contamination. Optimally we'd like to work with one NDE system that can detect and measure all of the above variables. Unfortunately, one such system or technology does not presently exist.

The following NDE test systems appear to be promising candidates at this time for the inspection of space structures: acoustic emission, strain gauges, thermocouples, pressure sensors, accelerometers, ultrasonics, holography, radiography (computed tomography, laminography, digital), eddy current, thermography, dye penetrant, and nuclear magnetic resonance. Although the list is long, it is by no means all inclusive; it just includes the NDE techniques that were most frequently mentioned or are being applied to space structure inspection. Some discussion on the most promising candidates follows.

Thermography is an infrared scanning system which provides a color map of the target surface which shows constant temperature contours. The output can be digitized which makes the data easier to analyze. The advantages to this technique are full structure evaluation, coupling transducers are not needed, and the NDE signal can be transmitted to a remote work site. Present disadvantages are the bulky systems, they may require robotics for unmanned vehicles, and multiple camera networking. Laser based ultrasonics is a very promising NDE technology for both earth and space-based operations. The advantages include portable systems, rapid structural inspection, and no couplants are required. The fact that no couplants are required is probably the largest advantage to this type of system. Coupling, though readily achievable in the earth's atmosphere, is a major problem in space, not to mention the fact that the couplant can contaminate local space surfaces. There are some disadvantages to laser based ultrasonics. One question that always arises when lasers are in operation is the safety of having a laser beam focused at various components, especially if the structure being inspected is a manned vehicle. The sensitivity level using lasers is also a question, along with the main factor that this is still a laboratory type of technology. Acoustic emission (AE) is another NDE technique that holds great promise for both earth and space-based operations. This technique is especially good for composite structures because it listens to the structure as it is stressed, and can detect and locate growing cracks. Also, transducers can be embedded into a structure. Some of the disadvantages that must be overcome are that AE requires the use of many transducers, extraneous noise overwhelms the target signal, and the technology is not in extensive field use yet.

Computed tomography (CT) is an NDE technique which uses an x-ray irradiation source and solid state x-ray detectors. These detectors contain scintillating crystals which convert the transmitted x-rays into visible light signals which are then converted into digital electronic signals by photodiodes. After scanning is complete, the digital data are then stored for future use or immediately manipulated to present an enhanced view of the object which is far superior to that provided by film x-ray. This view, or tomograph, is a cross-sectional mapping of the relative densities of discrete volume elements (voxels) within an object plane. If the CT machine is properly configured, the size of the voxels may be easily varied to suit the particular inspection needs.

CT technology is rapidly being applied to the inspection of solid rocket components and replacing the prevalent tangential x-ray technique described earlier. Unlike tangential x-ray, CT provides quantitative data on the inspected component, and the numbers are directly relatable to density. Should densities be related to mechanical properties, these can be fed into a thermal-structural analysis code to predict part performance, allowing an objective part evaluation. Due to recent advances in electronics and computer technology, CT is becoming an affordable quantitative inspection technique capable of inspecting thin-walled objects. Figure 3 shows an artist's conception of a space-based CT system which consists of the same type of components as an earth-based system, except that it has its own guidance and control system and a separate data transmission station. CT generates an enormous amount of data and would require rapid data transmission to a central analysis space or earth-based station. Perhaps the ultimate use of a CT system in space would be to inspect space structures and rocket com-

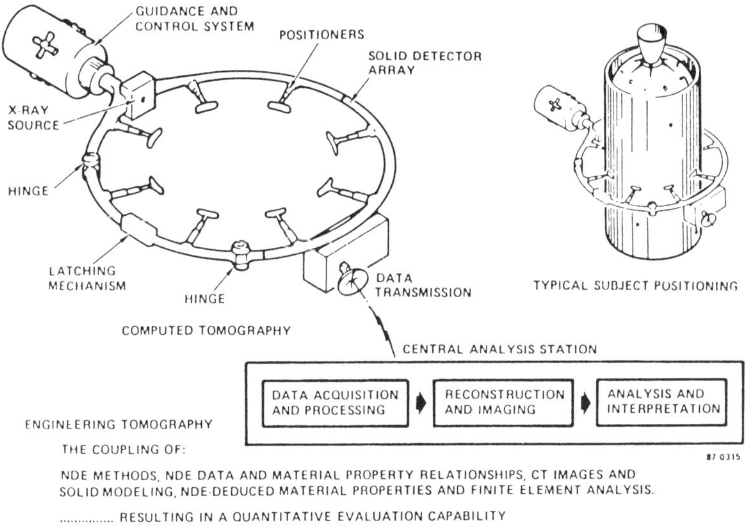

Courtesy of PDA Engineering, Newport Beach, CA

FIGURE 3- Computed tomography in space.

ponents prior to space launch. It is unknown what the extended space environment storage has done to a rocket motor. By CT inspection of the motor from inception to just before launch, an established data base would be available with which to compare the space-based inspection. This would allow a solid engineering judgement on the motor's integrity prior to launching. The inset in fig. 3 shows what a solid rocket motor inspection might look like in space.

Due to the high precision requirements of space-based systems such as space-based radar (SBR) and space-based laser (SBL) development programs are underway to embed sensors and actuators into composite structures. Figure 4 shows an artist's conception of a composite strut with embedded sensors and actuators. The main emphasis of the embedded control systems is for the dynamics and control of large space structures, but a side benefit of this technology is that the control systems can also be used for non-destructive health monitoring. There is also the potential that the same sensors can be used to perform both the control and health monitoring

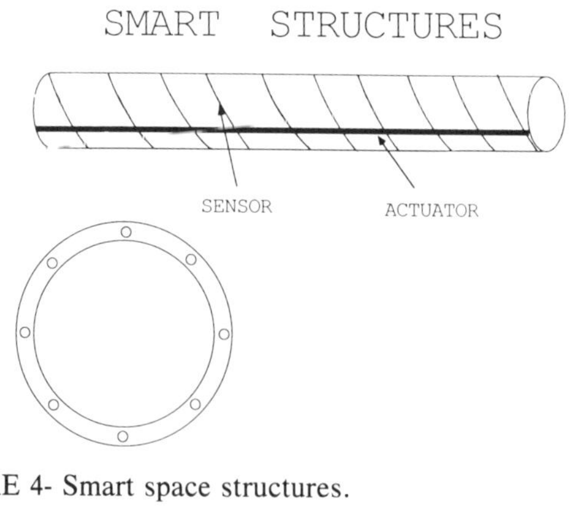

FIGURE 4- Smart space structures.

functions. The embedded control and health monitoring system would act like the human body's central nervous system. Work is presently being performed with fiber optics and piezo ceramics being embedded into composite panels and struts. The fiber optics, which are sensitive to motion, compression, and impact damage to the composite materials are being used as sensors, and the piezo ceramics are being used as actuators. Using commercial off-the-shelf fibers and stripping their coatings, the fibers are embedded into the laminate structures. Preliminary data shows that composite structures with the embedded control systems lose between 0 to 20 percent in strength. Work in this area will be ongoing at the AFAL over the next few years, with efforts focusing on laminate and filament winding construction, processing, and testing of space structures with embedded control and health monitoring system.

Three additional areas requiring attention in the development scheme of NDE for space are NDE equipment life cycle requirements, data management, and defect effects. Although these areas do not deal with the theoretical aspects on NDE for space, they have a significant impact on the usability of NDE in space. Current life cycle designs for space structures target a 20 year mission life. This doesn't necessarily mean that the NDE space inspection system should also have a 20 year life span. These NDE systems might be fabricated for servicing in space, or be returned to earth for refurbishment. Systems built into the space structures would have to meet the structural design and life requirements of the space structure, or would have to be one of many redundant systems on the structure. It is most desirable to develop the NDE system with a long life cycle to reduce down time and achieve low cost inspections.

The second area of concern is NDE data management. A tremendous amount of NDE data would be generated.

Such data will have to be catalogued, stored and reviewed at the depot level. This is not a small task since a space structure will undergo many NDE inspections. The data management function is the backbone of an effective NDE program, and efficient methods of reviewing NDE data need developing. The timeliness of data review is critical to the effectiveness of any NDE program. Hence, technology will have to be developed for all NDE inspection techniques which will yield the data to automated inspection. This then leads to a third area of concern which is defect effects. What does the NDE data mean to the user? How much usable life is left in a structure, or how much repair must one perform on a structure? This is not a small task to undertake. Experiments need to be performed which simulates the space environments a particular structure will see, and evaluate the structure's performance against an NDE signature. This type of data can then be used in a structural analysis code to predict system life. It's going to be a long road to achieve this goal.

SUMMARY

In summary, a lot of exciting work lies ahead for the NDE community with new challenges to be addressed daily. Due to the high cost of space structures, their mission profiles, and component life cycles, NDE becomes a necessity. NDE needs to be introduced at the very beginning of a structure's fabrication and followed until it is retired. Work needs to begin now so that the proper NDE technologies can be in place when required for use.

Evaluation of NDE Methods in the Space Environment

Glenn M. Light, Richard A. Cervantes, Victoria G. Muller, and Stephen N. Rowland

Southwest Research Institute,
6220 Culebra Road,
San Antonio, Texas 78284

ABSTRACT

Nondestructive evaluation (NDE) methods have long been applied in the earth environment to ensure the safe operation of critical components and facilities. By the mid-1990s, the United States is scheduled to have an operational space station in earth's orbit. This large complex structure will be subjected to extreme temperatures ($-250°F$ to $+350°F$), radiation, and low pressure (vacuum) atmosphere (10^{-9} torr). The space station and other spaceborne structures will need to be inspected prior to and during service to ensure their safe operation. Nondestructive evaluation methods must be capable of being applied to these structures in the space environment.

Four NDE methods (ultrasonics, acoustic emission, eddy current, and radiography) were evaluated in simulated space environment conditions. Transducers, couplants, probes, cables, and materials were tested at low and high temperatures, high vacuum, and low level radiation. Test plans developed to evaluate each NDE method are described, and the results of the tests, conclusions concerning applicability in the space environment, and recommendations are discussed.

INTRODUCTION

Nondestructive evaluation (NDE) methods have long been applied in the earth environment to ensure the safe operation of critical components and facilities. By the mid-1990s, the United States is scheduled to have an operational space station in Earth's orbit. This space station will be a large complex structure assembled in space from components delivered by the space shuttle and other delivery systems. In space, this structure will be subjected to extreme temperatures, radiation, and low pressure atmosphere. As with most fabricated structures which must support life, the space station and other spaceborne structures must be inspected prior to and during service to ensure their safe operation. To provide this type of information, nondestructive evaluation methods must be capable of being applied to these structures in the space environment.

BACKGROUND

Most of the NDE methods have been developed for use in earth environmental conditions. Some NDE inspection sensors have been developed to be used in high temperature (up to 500°F) and high pressure (such as experienced in down-hole oil well conditions) environments. However, no environment to date has posed the extremes that exist in space. The space environment provides an extreme temperature range from −250°F to +350°F, a background radiation level of 100 mR, and high vacuum (10^{-9} torr).

To determine the applicability of current NDE methods for the space environment, Southwest Research Institute internally funded a feasibility study. The main objectives of this Internal Research program were to identify and evaluate potential problems with the application of four NDE methods; namely, ultrasonics, acoustic emission, eddy current, and radiography, in a simulated space environment and make recommendations for future developments.

TESTING AND RESULTS

Each NDE method was addressed separately in terms of application in the space environment. The probes and application of the probes for each method were tested in the appropriate simulated environments. The ultrasonic (UT) method was tested in the temperature environment, in the vacuum environment, and in the combined temperature/vacuum environment. The acoustic emission (AE) method and the eddy current (EC) method were tested in the same manner as the UT. The radiographic method was tested in the same manner as the UT, and additional tests were conducted in a 100 mR/hour radiation field. The testing and results of the tests are described below for each method.

Ultrasonic Testing

Perhaps the most widely used nondestructive evaluation method is ultrasonics. The major applications of ultrasonics for NDE include flaw detection, thickness measurement, and characterization of metallurgical structure. Among the advantages of this method are its ability to penetrate to substantial depths in many materials, its sensitivity in the detection of small flaws, and its capability for rapid scanning. Principal disadvantages include the need for skilled technicians for proper application and interpretation and the need for essentially good, direct mechanical contact (or coupling) with the test part, a requirement which is often difficult to achieve in practice.

In practice, the ultrasound energy is produced by a piezoelectric crystal (transducer) and introduced into the test part by direct mechanical coupling. To optimize the sound transmission from the transducer into the test material, couplants, typically viscous fluids, are used at the point of contact. It is these components, the ultrasonic transducer and the couplants, which suffer from exposure to harsh environmental conditions. Cables and connectors are also susceptible to harsh conditions; however, the state of the art in these areas is such that products and techniques are available to tolerate in-space conditions. Thus, only couplants and transducers were tested to determine the problems associated with their use in the space environment.

Couplant Testing. For in-space use, couplants must withstand severe temperature extremes and acute vacuum conditions. Since their viscosity and vapor pressure are strongly temperature dependent, many standard substances will not survive the exposure. They are susceptible to evaporation, thickening, and/or burning, rendering many totally useless. Some commercial couplants are available for limited use in a high temperature environment; however, very little experience exists in the application of ultrasonic NDE techniques at low temperatures or in vacuum conditions.

Several couplant materials from various manufacturers were tested at elevated temperatures, at low temperatures, in a vacuum, and in combined conditions to determine their performance capabilities. Most of these couplants were capable of withstanding the high temperature for only short periods of time, typically 15 seconds to a few minutes. As a survivability test, these substances were placed in individual trays within an oven and the temperature was slowly raised to 350°F. The couplants were then examined after this exposure. The results showed a range of performance from evaporation and outgassing to thickening and discoloration. As a further and more severe test, new samples of the same couplants were also subjected to simultaneous conditions of high temperature and vacuum conditions (10^{-6} torr). The results of this test were similar to the previous test; however, some of the samples experienced more thickening and one sample became crystallized.

Two couplants were selected for further testing with active ultrasonic transducers in high temperature and vacuum conditions. In these tests, a transducer was placed on an aluminum block containing a known reflector. The transducer, with the test couplant, was secured to the block, placed in the heated vacuum chamber, and connected to a UT instrument through a cable port. The reflector signal amplitude was monitored before, during, and after the test. The results showed a decreasing signal amplitude as the temperature reached a maximum of 350°F. Upon returning to ambient conditions, the signal amplitude did not return to the original level. However, this observation could not be isolated solely to the couplant or transducer. A post-test verification of the transducer and reflector signal amplitude (with new couplant) showed that the transducer had apparently been internally affected by the exposure and was no longer performing in its original condition. The two couplants were still viscous, showing little or no effect as a result of the exposure, and performed well as ultrasonic couplants in post-exposure verification tests. Therefore, these couplants satisfied the test requirements.

The low temperature extremes of space were simulated in a liquid nitrogen cooled chamber and bath. Couplant samples were slowly cooled by the liquid nitrogen boil-off and eventually by immersion into liquid nitrogen. Sample temperature was monitored with a thermocouple. Several of the couplants that performed well in the high temperature vacuum tests were chosen for these cold temperature tests. As the temperature was lowered, each couplant was checked to determine if solidification had taken place. The results showed that all couplants tested became more viscous and had solidified by the time they reached approximately −50°F. Nonetheless, the temperature was lowered to liquid nitrogen temperature (−300°F) to complete the test.

The results of these environmental tests on couplants indicated that couplant selection for in-space use must be carefully considered and tested. Commercially available couplants have been shown to survive and perform reasonably well in high temperature and vacuum conditions. However, none of the couplants tested retained viscous properties at the low temperatures.

Ultrasonic Transducer Testing. Conventional ultrasonic transducers are normally used in ambient conditions or at least in conditions tolerable by their human operators. In the vast majority of situations, this is not a restriction and actually simplifies design and enhances the performance of ultrasonic transducers. While some applications of terrestrial ultrasonic NDT require special preparation, no earth-bound tests pose the wide extremes encountered in the space environment. As the single most critical element in ultrasonic NDT, transducers are also most susceptible to environmental conditions. Thus, to be considered for use in space, transducer technology must be thoroughly studied to determine its vulnerabilities to such extreme conditions. In this study, tests were conducted to begin to identify the transducer characteristics that are significantly affected by wide temperature variations, low pressure (vacuum), and the combined conditions.

A typical ultrasonic transducer is constructed of materials of vastly different physical characteristics. Temperature variations result in expansion and contraction of the individual components of a transducer. This is not usually a concern in near-ambient conditions since relative temperature variations are small and only minute changes result. In the space environment, however, materials may experience near-cryogenic temperatures (−250°F) at one instance, and then be exposed to direct solar radiation which may raise surface temperatures to 350°F or more. These temperature variations are significant to transducer performance and survivability. Tests in this environment were conducted on several conventional ultrasonic transducers. These results showed that thermal expansion of the transducer elements can be serious enough to damage the crystal and render it useless. Thermal insulation on transducers may reduce the effects of exposure to the wide temperature variations; however, it is obvious that standard transducers are affected to a significant degree by the temperature environment.

Further tests were carried out on active transducers. The signal was monitored with an ultrasonic instrument, and the pretest conditions and signal amplitude were recorded for the reference. Beginning from an ambient temperature of 75°F, the sample and transducer temperatures were slowly raised to 350°F. At specific times during the test, the temperature and reflector signal amplitude were recorded. The data showed a decreasing signal amplitude as the temperature increased, as illustrated in fig. 1. At approximately 300°F, 50°F below the upper limit of the test, the signal became intermittent and had lost over 12 dB of amplitude. As the temperature gradually returned to normal, the signal amplitude did not recover to its original condition. After the test, the transducer was sectioned, revealing separation of the piezoelectric crystal from the thin epoxy face and separation between the backing material and crystal.

Low temperature tests were also performed on a standard UT transducer. In this test, a transducer was again secured to the aluminum block and placed in a liquid-nitrogen-cooled chamber. The reflector signal was monitored throughout the duration of the test. The signal am-

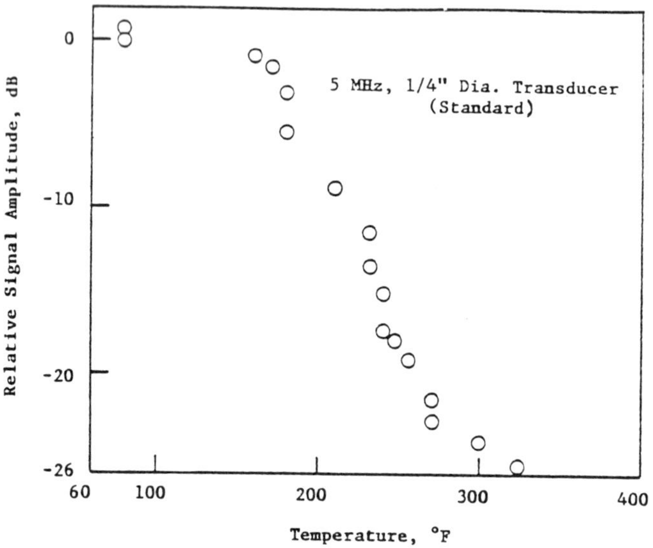

FIGURE 1- Plot of the conventional transducer performance at high temperature. At 325°F the signal amplitude drops by approximately 24 dB and has become intermittent.

plitude dropped 4dB after reaching only 200°F and continued to drop as the temperature cooled further. The amplitude of the reflected notch signal was varying but was still discernible (– 10dB) at – 80°F although the baseline noise had increased significantly. At – 100°F the amplitude rose to nearly normal and then quickly dropped into the noise (– 15dB) at about – 115°F. The test was continued down to a temperature of – 280°F before slowly warming back to ambient conditions. At ambient conditions, the signal amplitude recovered fully to its original level. It was apparent that couplant solidification was a problem which could have contributed to the effects observed during the cold temperature test.

Piezoelectric Characteristics. The characteristics of the piezoelectric crystals have been shown to be temperature-dependent[1,2] above room temperature. During this study, the piezoelectric characteristics of lead metaniobate ceramics were studied as a function of temperature over the range of −250°F to +350°F. The characteristics monitored during the tests were the crystal frequency, the crystal capacitance, and the piezoelectric coupling factor (which is the amount of electrical energy converted into mechanical energy). These data were obtained by using a special circuit designed to pulse the crystal and monitor the impulse voltage and the ring down of the voltage across a capacitor.[3]

Plots for frequency, capacitance, and piezoelectric coupling factors showed in general that the frequency varied approximately 40% over the temperature range. At the low temperature, the frequency became higher by approximately 30% and at the high temperature the frequency was approximately 10% lower than nominal. The piezoelectric coupling factor varied approximately 20% over the temperature range. The data showed that the coupling factor increased monotonically over the range, being lower at the low temperature and higher at the high temperature. The capacitance also varied over the temperature range by approximately 20%. The minimum capacitance was observed at the lower temperature and the higher capacitance at the high temperature.

In an attempt to overcome these low temperature problems, a different type of transducer was designed, fabricated, and tested. The results of the high and low temperature tests showed that this transducer could survive the temperature extremes. The data shown in Figure 2 illustrate how the new transducer performed at high temperatures. This transducer was also tested at low temperature. The data showed a decrease in notch reflector signal amplitude of 4dB from +78°F to −256°F.

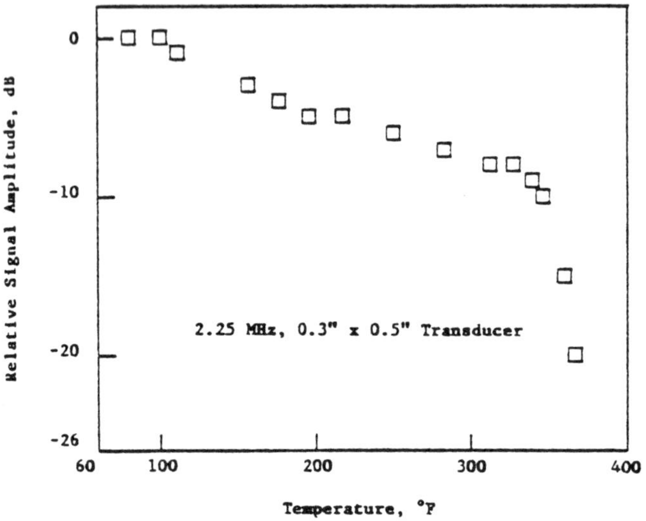

FIGURE 2- Performance of prototype improved transducer at high temperature. The gradual drop in signal amplitude allows use of the transducer up to 350°F.

Acoustic Emission Testing

Most materials used in structures (metals, composites, concrete, etc.) are known to emit acoustic energy when stressed to a certain level. This characteristic is the basis of the acoustic emission (AE) nondestructive evaluation technique. Piezoelectric transducers, very similar to those described earlier in the ultrasonics section, are placed on a structure in order to passively "listen" for acoustic emissions generated by the material. The presence of flaws, such as cracks, voids, and other discontinuities, releases acoustic emissions with an applied stress.

The advantages of the AE technique stem from the fact that it is essentially a nonlocalized technique. That is, it is not necessary for the receiving transducer to be very close to the source of the emissions. However, with only a few transducers, a large area can be monitored and flaws can be located. Also, the AE technique is well suited for continuous inservice monitoring of stressed structures. The disadvantages of the AE technique are due to its sensitivity to emissions from many noise sources. The typical background noise of a structure must be well understood in order to allow flaw noise discrimination. Thus, care must be taken to minimize noise emissions from nonrelevant sources.

For use in the space environment, AE transducers must be capable of operating at maximum sensitivity in spite of the extreme environmental conditions. Furthermore, with space structures exposed to wide temperature variations, background noise levels are expected to be significant. A clear understanding of the background noise conditions will be required for in-space AE monitoring.

All of the AE tests performed during this study were conducted using a microprocessor-controlled AE system which could record signals from up to four AE sensors with independent control of gain and detection threshold. An additional in-line preamplifier was used in the testing to enhance low-level signal detection. Key parameters of AE events such as event time, count, amplitude, energy, duration, and others were plotted as distribution functions in graphical form for clear analysis of the test data.

Three different types of AE sensors were evaluated during this study. For the contact sensors, a high temperature couplant was used to provide good sound transmission. During all of the tests, the sensors were connected to preamplifiers and then connected to the AE system. The three sensors were rated for service up to a temperature of 437°F, but were not given a specific rating at the lower temperatures. During this study, the transducers were tested after each test to check for any loss of sensitivity resulting from the cold temperature exposure cycle. The calibration was done by using a reference transducer as a source coupled with 90 wt. oil to the test transducer. An AE simulator was used to pulse the reference transducer while both the reference signal and the detected signal were monitored on an oscilloscope. These results were recorded to document pre-test transducer performance for later comparison with an identical post-exposure test.

For most of the linear location tests, a simple pencil lead-break calibration was performed on the test sample prior to each test to identify the location of the AE transducers and, where applicable, the location of the flaw.

The high temperature tests of AE transducers consisted of suspending the transducer inside an oven and raising the temperature to 300°F. As the oven temperature was raised, any acoustic events were recorded along with the test time and stored on disk. After the test, the data were displayed showing AE events as a function of test time. The results showed a very low number of recorded emissions during the full test period.

Since the previous testing showed that transducer noise was not a serious problem at high temperature, another test was conducted to determine how much noise was generated by transducer-to-part contact and by material noise due to high temperature cycling. A transducer was mounted on an unflawed 6061-T6 aluminum sample (6″ × 1″ × $\frac{1}{4}$″) and exposed to 300°F. Again, data were recorded as the temperature slowly increased. These results showed a significantly greater number of acoustic events than with the transducer alone. Thus, contact area and material noise will be a significant source of emissions in high temperature monitoring of structures.

Cold temperature testing of the acoustic emission transducers and techniques was conducted in the environmental chamber cooled by liquid nitrogen. Since the AE transducers were not rated for cold temperature operation, the transducers were tested before and after each test to verify their functionality. The AE simulator was used to generate a repeatable pulse which was monitored on an oscilloscope. The transducer was placed in contact with the test transducer and both the transmitted and the received pulses were recorded. In the post-exposure verification, changes in transducer performance could be detected as distortion of the received waveform or by a reduction in its amplitude.

To determine how much noise was generated by an AE sensor during exposure to extreme cold temperatures, a sensor was exposed to $-250°F$. Relatively few emissions were recorded and sensor failure did not occur. These results indicated that contact and material noise were not as prominent as in the high temperature tests.

Eddy Current Testing

Eddy current (EC) testing is widely used[4,5] in the aerospace industry for examination of components before service and during aircraft maintenance periods. In the space environment, EC can be used to determine the integrity of plated and coated surfaces and to locate and measure discontinuities such as cracks. These tasks require both differential and absolute measurements of the material properties. Differential sensors are generally used for flaw detection where high sensitivity is required over a small area. Absolute sensors can also be used to detect flaws, but are generally used for material property measurement or thickness measurement.

Eddy current relies on the ability to measure the impedance or differences in impedance in the sensor coil of an eddy current probe due to changes in a material or in the relationship between the sensor and the material. The impedance of an eddy current sensor has a resistive component (magnitude) and a reactive component (phase). The magnitude of the resistance of the wire in the sensor and the cable is a fucntion of temperature because the resistivity and the dimensions of the wire are temperature-dependent. The inductive reactance (or phase) is also affected by temperature since it is a function of the physical dimensions of the coil and the eddy current distribution in the sample.

The effects of high temperature are also well documented.[4,6] The biggest challenges, however, are compensation for changes in the sensor and the materials at low temperatures. Therefore, this study focused on temperature effects in the range of +100 to −250°F. The effects of radiation are well documented in the literature for radiation fields up to 1×10^6 Rem.[7] Since the radiation levels in the space environment are approximately 100 mR/hr, the effects on the EC method are negligible.

Absolute Probes. The absolute probe uses a single eddy current sensing coil to sense changes in material properties or dimensions. The impedance of the sensor is the sum of the inductive reactance and the resistance. If a flaw is present, there will be a variation in impedance as the sensor passes over the area containing the flaw. This variation is a small percentage of the total impedance for small flaws; therefore, variations in sensor properties are significant in a test using this type of sensor.

At each temperature the frequency was varied from 100 Hz to 7 MHz and a set of data consisting of the magnitude and phase of the impedance at each frequency was recorded. Data were stored on disk for postprocessing. An environmental chamber was utilized to control the temperature of the sensor and any samples present in the test.

Three types of experimental data were taken for the absolute probe experiments: sensor in air, sensor on 2024 T6 aluminum block, and sensor with block and flaw. The sensors used in this study were designed for operation at temperatures up to 500°F. However, this design had never been tested at the low temperatures where the ductility of all the materials in the sensor decreases. The coil forms were made of Teflon and the coil windings were copper magnet wire #38 AWG with a polyurethane coating. A ceramic wear face was used to act as one side of the coil form to provide protection from abrasion. The coil, ceramic wear face, and forms were embedded in a high temperature, filled epoxy.

Temperature effects on the structure of the sensor are created by different expansion coefficients. In this design, the greatest mismatch occurred between the Teflon coil forms and the other parts of the sensor. After a series of tests, the epoxy separated from the Teflon and the ceramic wear face developed numerous cracks. The sensor still functioned, however, and showed little degradation in performance.

Differential Probes. Differential eddy current probes are, in effect, two absolute coils operating in a bridge circuit so that their output is zero when both coils are affected in the same way. This principle makes this type of sensor more sensitive to small flaws than the absolute sensor and less sensitive to changes in parameters that affect both coils in the same way. These effects can be reduced if both coils in the differential pair have similar electrical properties.

The experimental setup for evaluation of the flaw response as a function of temperature was similar to the method used for absolute sensors. An MIZ-17 eddy current instrument was used along with a differential sensor and a smaller chamber which allowed scanning of the sensor over three flaws 0.006″ wide × 0.010″ deep, 0.006″ wide × 0.020″ deep, and 0.015″ wide × 0.030″ deep in a plate parallel to the other notches in an aluminum block. The sensor was operated at two frequencies simultaneously (0.5 MHz and 2 MHz) and the output was recorded by the computer as the sensors were scanned over the flaws (as shown in fig. 3).

A quantitative comparison of the effects of temperature on flaw response was made by comparing the response from the 0.020-inch deep notch and at 0.5 and 2 MHz every 50°F. The experimental data for temperatures from −250°F to 0°F were compared to the model data in the same temperature range. In fig. 3, the response from the 0.020-inch deep notch is shown at the two frequencies.

If differential sensors are used, the effects of temperature on the EC signals can be reduced. Effects from the change in liftoff will continue to be seen, but now effects of changes in the material properties of the block can be seen. In the data from the differential sensor, effects of liftoff and material properties combine to produce a decrease in the magnitude and phase angle (effective decrease in depth) as the temperature decreases. The effect of the liftoff only should be an incrase in the magnitude of the flaw response because the sensor coil is closer to the flaw. However, as the temperature decreases in the block, the conductivity increases, which decreases the skin depth and reduces the phase content of the eddy current distribution in the block affected by the flaw. Therefore, using differential sensors, the predominant effect is due to changes in material properties rather than the liftoff change. Detection of flaws was not greatly affected.

Radiographic Testing

A major consideration for successful in-space application of the radiographic method is the susceptibility of the film to the harsh environmental conditions. The effects of exposure of standard radiographic film to high and low tem-

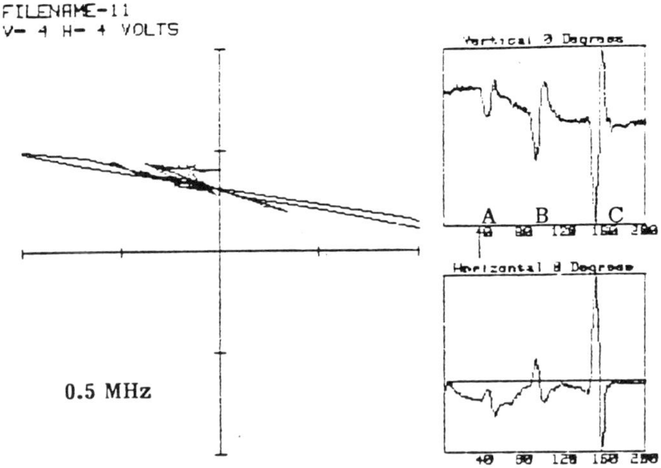

Impedance Display

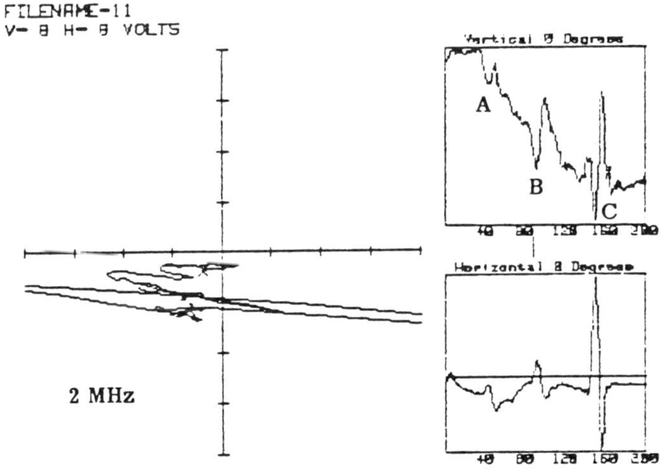

Impedance Display

FIGURE 3- Response at 0.5 and 2.0 MHz on impedance display at temperature of $-253°F$.

peratures, vacuum conditions, and simulated background radiation were studied.

The chemical nature of radiographic film makes it sensitive to temperature variations. For use in the space environment, radiographic film must be protected from or be capable of withstanding a temperature range of −250°F to +350°F. To determine the temperature sensitivity of the film, a series of tests were conducted in which radiographic film was exposed to x-rays while in the environmental chamber at a controlled temperature. Two types of standard radiographic film were studied: type AA film, a medium speed industrial film, and type T film, a slower speed film with greater resolution. Most of the testing, however, was conducted with the type AA film. Common fiberboard or plastic film cassettes without any intensifying screens were used during this study. As a reference subject, an aluminum step wedige calibration block was used to produce a film image with sufficient film density range. The calibration block consisted of six machined steps with thicknesses of $\frac{1}{8}''$, $\frac{1}{4}''$, $\frac{3}{8}''$, $\frac{1}{2}''$, $\frac{3}{4}''$, and 1″. The radiographic technique chosen employed 110KV x-rays at 4 milliamp current with a total exposure time of 2 minutes. These settings resulted in radiographs at ambient conditions with measured film density levels of 1.5 to 4. Two films were exposed simultaneously. One film was placed inside the environmental chamber for exposure to extreme temperature conditions, while the second film was placed outside the chamber as a reference (as shown in fig. 5). In this fashion, any variations in exposure from test to test would be indicated in the measured density values of the reference films.

Several tests were conducted at ambient conditions to establish the behavior of the film under normal circumstances. Within the chamber, the test film was exposed to a temperature environment controlled by either liquid nitrogen boiloff (for low temperature tests) or by heating

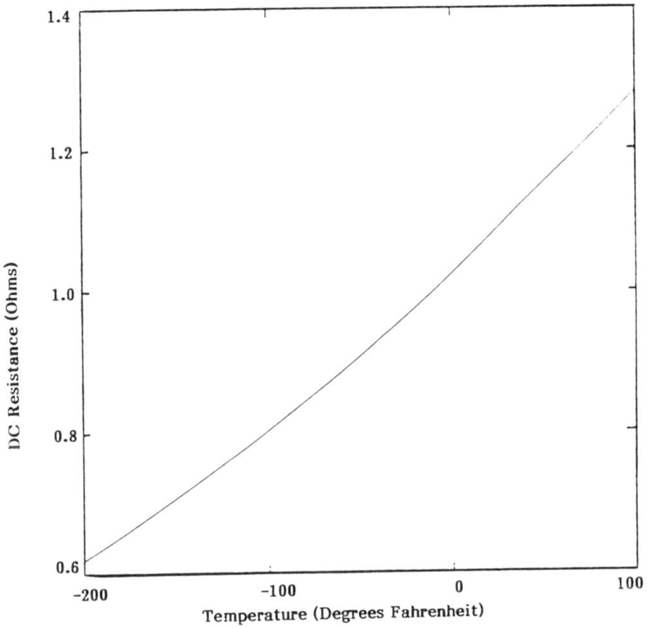

FIGURE 4- DC resistance curve resulting from fitting several second order polynomials over temperature range from $-200°$ to $+100°$.

elements (for low temperature tests). A series of tests were then conducted in which the radiographic film was subjected to low temperatures with sufficient time to ensure film temperature stabilization (about 20–30 minutes once the chamber temperature had achieved the desired level). Cold temperature tests were conducted at 10°F, −55°F, −125°F, −200°F, and −275°F. Each time, a reference film outside the chamber and at ambient conditions was taken simultaneously. For the high temperature tests, heating elements were placed inside the chamber

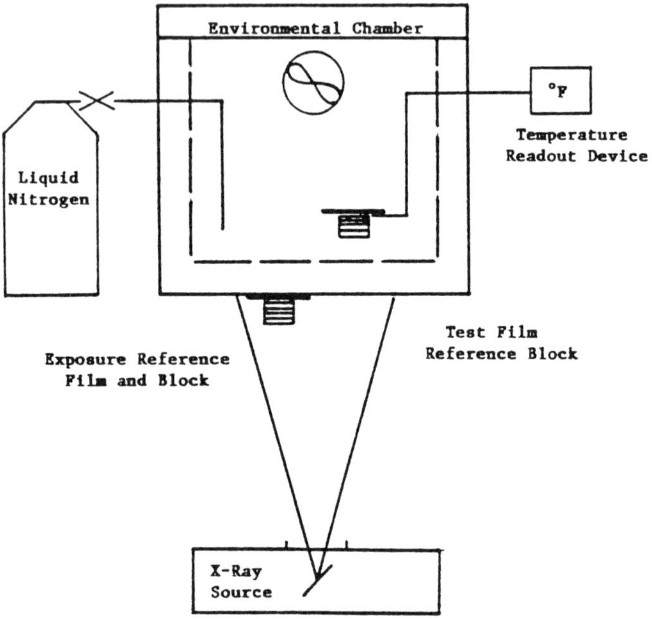

FIGURE 5- Illustration of experimental setup for testing of radiographic film at high and low temperatures. A reference film was placed outside the chamber to account for slight variations in exposure from test to test.

and were controlled by a standard temperature controller. In the high temperature tests, temperatures of +120°F, +170°F, +220°F, and +270°F were achieved.

The effects of temperature on film sensitivity were determined by measuring the film image density at each of the steps depicted on the radiograph. Five readings were taken at each step location and the values were averaged. The test results indicate that film sensitivity is affected by temperature. As the temperature decreased, the film optical density values also decreased. However, even at the lowest temperature (−275°F), acceptable radiographs were still achieved. The high temperature tests showed significantly less tolerance to severe environmental exposure. At a temperature of 120°F, a slight change in optical density values as compared to an ambient condition exposure was observed. At 170°F, the measured film optical density values were more scattered and deviated significantly from the normal characteristic film curve. Exposures taken above 170°F were completely un-

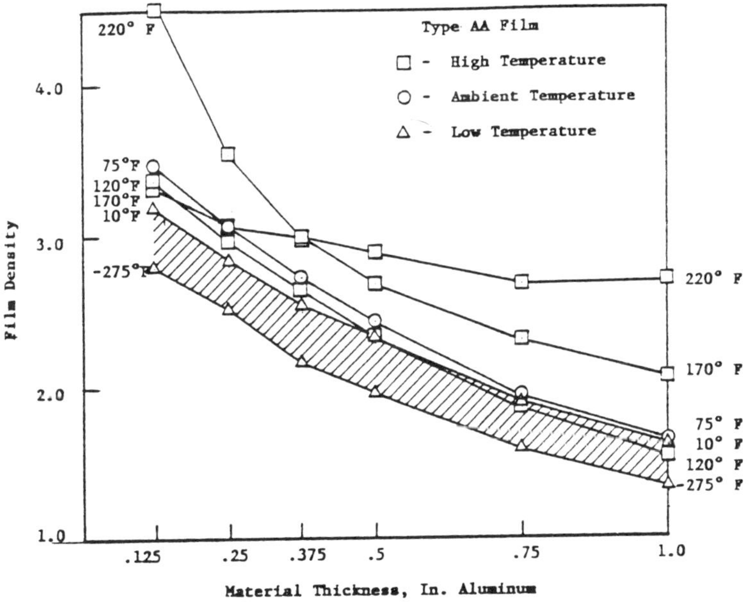

FIGURE 6- Experimental results of the effects of temperature on radiographic film density. Radiographic film is highly sensitive to temperatures above 120°F. The shaded region includes several tests at temperatures between 10°F and −275°F.

acceptable. A graph of the normalized film density values for both high and low temperatures is shown in fig. 6.

A previous literature search[8] revealed some limited testing of radiographic film exposed to vacuum conditions at ambient temperatures. The reported data showed no significant change in film quality as a result of the vacuum exposure. These results were verified during our study.

In the environment of space, the background radiation environment is estimated to be in the range of 100 mR/hour, consisting of gamma rays. For long-term exposure, this level of radiation could result in fogging or premature exposure of radiographic films. To simulate this condition, type AA film specimens were placed in front of a gamma ray (iridium-192 source with a strength of 38 curies) isotopic source. To simulate different total exposures, the film specimens were placed at specific distances from the source during an exposure time of $\frac{1}{2}$ hour. The simulated total exposures were 100 mR, 500 mR, 1000 mR, and 1500 mR. A penetrameter was placed on the front surface of each cassette in order to produce an image which might be discernible.

CONCLUSIONS

The data acquired during this program showed that all four of the NDE methods studied were very susceptible to one or more of the environmental conditions found in space. For the ultrasonic, acoustic emission, and eddy current methods, the cold environment had the most effect, while the cold had little effect on radiography. High temperature caused problems for all methods. The high vacuum affected ultrasonics because of the requirement to couple the ultrasonic energy into a part, but did not adversely affect the other methods. The radiation environment caused problems for film radiography but no problems at all for the other methods.

References

1. Kulcsar, Frank. "Electromechanical Properties of Lead Titanate Zirconate Ceramics Modified with Certain Three- or Five-Valent Additions." Journal of the American Ceramic Society, Vol. 42, No. 7, July 1959, pp. 343–349.
2. Jaffe, H. and Berlincourt, D. A. "Piezoelectric Transducer Materials." Proceedings of the IEEE, Vol. 53, No. 10, October 1965, pp. 1372–1386.
3. Stuetzer, O. M. *Transactions on Sonics and Ultrasonics*, Vol. SU-15, No. 1, January 1968, pp. 13–17.
4. Reed, Richard P. and Clark, Alan F. "Materials at Low Temperature." American Society of Metals, Metals Park, Ohio, 44073. Canner Publication Services Inc., 1983, pp. 164–166.
5. Ariel, A., Bar-Cohen, Y., and Blas, A. "The Effect of High Temperature Oxidation on the Conductivity of Ti-6A1-4V Alloys." Materials Evaluation, May 1978, pp. 62–64.
6. Linanov, I. A., Abinomov, M. A., and Denisov, V. A. "Electromagnetic Transducer with Temperature Compensating Shield." Translated from Defektoskopiya, No. 1, January 1982, pp. 61–64.
7. Gross, B., Biacometti, J., and Leal-Ferreira, G. F. "Corona Method for Investigation of Charge Storage and Transport in Dielectrics." IEEE 1981 Annual Report on Conference on Electrical Insulation and Dielectric Phenomena, October 26–28, 1981, Whitehaven, Pennsylvania, pp. 39–44.
8. "Nondestructive Equipment Study." TRW Final Report RA-509T, TRW Space & Technology Group, Redondo Beach, California, January 1985.

Index